ADVANCED
MICROPROCESSORS

Vilas M. Ghodki

Satish J. Sharma

Trupti A. Dange

ISBN-10: 1496086112
ISBN-13: 978-1496086112

DEDICATION

To our beloved

parents and teachers

CONTENTS

Acknowledgments   i

I   Introduction to 8086   1

II   Instruction set of 8086   25

III Assember directives   44

IV   Stack   59

V   Assembly language   73
     programming (ALP)

VI   Minimum mode of 8086   115

VII Architecure of 80286   129

VIII Architecure of 80386   149

IX   Architecure of 80486   164

ACKNOWLEDGMENTS

We greately acknowledge the contribution made by our teachers, friends and students in bringing out this book.

Dr Vilas M Ghodki

Dr Satish J Sharma

Mrs Trupti A Dange

CHAPTER I

INTRODUCTION TO 8086

KEY FEATURES OF 8086

1. HMOS technology
2. 16 bit microprocessor
3. 20 bit   address bus and 16 bit data bus lines [16 lines multiplexed address/data]
4. Total memory 2^{20} =1 Mbytes
5. It consists of more than 36,000 transistors
6. Clock frequency :5 MHZ, 8MHZ,10MHZ   for different versions
7. Incorporating pipeline technique (6 byte instruction pipe)
8. Architecture permits its usage in multiprocessing, multiprogramming ,time sharing system environment
9. Powerful set of instructions :136
10. + ,_ ,* ,/ and square root ,exponential etc can be performed using mathematical co-processor 8087
11. It requires only +5V power supply and current 360 mA

12. It operates in two modes minimum and maximum.

13. Provides 256 types of vectored software interrupts.

14. Supports 24

ARCHITECTURAL BLOCK DIAGRAM OF 8086

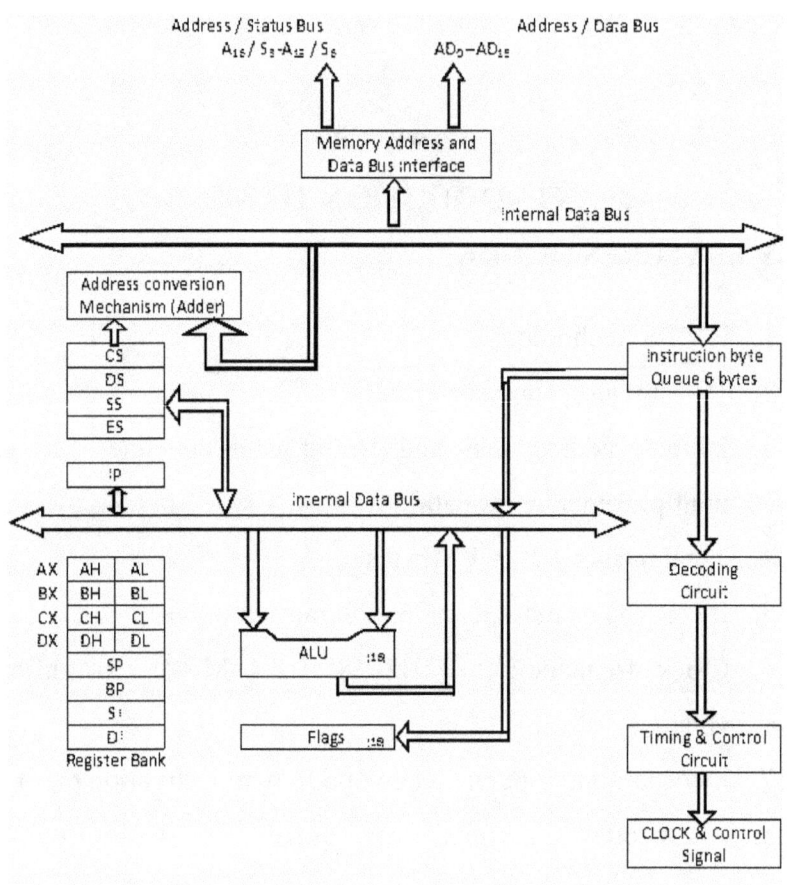

CS: Code segment

DS: Data segment

ES: Extra segment

SS: Stack segment

IP: Instruction Pointer

AX : Accumulator

BX: Base register

CX: Counter register

DX: Data register

SP: Stack Pointer

BP: Base Pointer

SI: Source Index

DI: Destination Index

BIU: Bus Interface Unit

EU: Execution Unit

Intel 8086 is 16 bit H-MOS microprocessor . It is 40 pin IC and uses +5V DC Power supply for its operation. 8086 uses 20 line address bus. It can directly address up to 2^{20} = 1 MB of memory address. It uses 16 line data bus.

The 20 lines of address bus operate in multiplexed mode. The 16 low order address bus lines are multiplexed with data ($AD_0 -$ AD_{15}) and 4 high order address bus lines are multiplexed with status signals ($A_{16}/S_3 - A_{19}/S_6$)

The complete architecture of 8086 can be divided in to two parts:

1} BIU: Bus Interface Unit

2}EU: Execution Unit

The BIU contains the circuit for physical address calculations and a pre -decoding instruction byte queue (6 byte long). This unit is responsible for establishing communications with external devices and peripherals including memory via the bus.

The EU executes the previously decoded instructions concurrently. It contains register set, ALU, timing and control unit etc.

PIPELINE/QUEUE:

The BIU along with EU forms a pipeline. In 8085, every time each and every instruction is first fetched, decoded and then executed. The processing time is saved in 8086. In 8086, next 6 instruction codes are pre-fetched and arrange in a queue called pre – decoded instruction byte queue. It has FIFO (First In First Out) structure and hence called pipelined architecture.

The instructions from the queue are taken for decoding sequentially and automatically next op-code is fetched simultaneously. Hence it saves the time and speed up the processing.

Physical address calculation:

The complete physical address which is 20 bit long is generated using segment and offset registers,

each 16 –bit long. For generating physical address, the contents of segment register called segment address is shifted left bit wise 4 times and to this result ,contents of offset register called offset address is added. For ex.

Segment address : 1005H

Offset address: 5555H

Segment address is 1005H is 0001 0000 0000 0101 B

It is shifted by 4 bit position to left side:

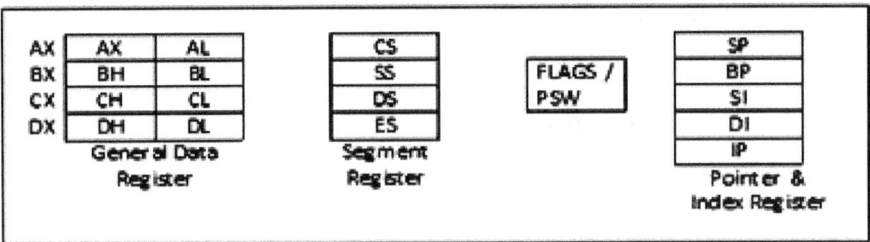

The physical address is 155A5H . Thus

(Physical address) =10H * (segment address) + (offset address)

REGISTER ORGANISATION OF 8086

AX	AX	AL		CS				SP
BX	BH	BL		SS		FLAGS		BP
CX	CH	CL		DS		/PSW		SI
DX	DH	DL		ES				DI
	General Data			Segment				IP
	Register			Register				Pointer &
								Index
								Register

Intel 8086 microprocessor contains 16 bit registers, grouped as follows:

1. General purpose register

2. Point and index register

3. Segment register

4. Instruction register

5. Status / flag register

1. General purpose register

There are four 16-bit general purpose registers AX, BX, CX & DX. Each of these 16 bit registers are further subdivided into two 8-bit registers as AL, AH, BL, BH, CL, CH, DL, DH. The letters L & H specify the lower and higher byte of a particular register.

Register AX servers as accumulator register BX servers as base register for computation of memory address. Register CX used as a counter register in case of multi iteration instruction. Register DX is used for memory addressing when data are transferred between I/O port & memory using certain I/O Instruction.

2. **Pointers and Index register**

The following four registers are the group of pointers and index registers

a) Stack pointer

b) Base pointer

c) Source index

d) Destination index

The pointers contains offset within the segment. Base pointer contains offset within the data segment. The function of Source index contain offset within stack segment. The function of source pointer is same as stack pointer in 8085.

The index registers are used as general purpose register as well as for offset storage in case of Indexed, base indexed & relative base indexed addressing modes. Register SI is used to store the offset of source data in data segment while register destination index is used to store the offset of destination in data or extra segment.

3. Segment register

There are four segment register

a. Code segment register

b. Data segment register

c. Stack segment register

d. Extra segment register

8086 addressess the segmented memory. The complete 1Megabytes memory is divided into 16-bit logical segments. Each segment, thus, contains 64 Kilobytes of memory. The segment register of 8086 acts as base register. The segment register point out the starting memory address of the currently used segments.

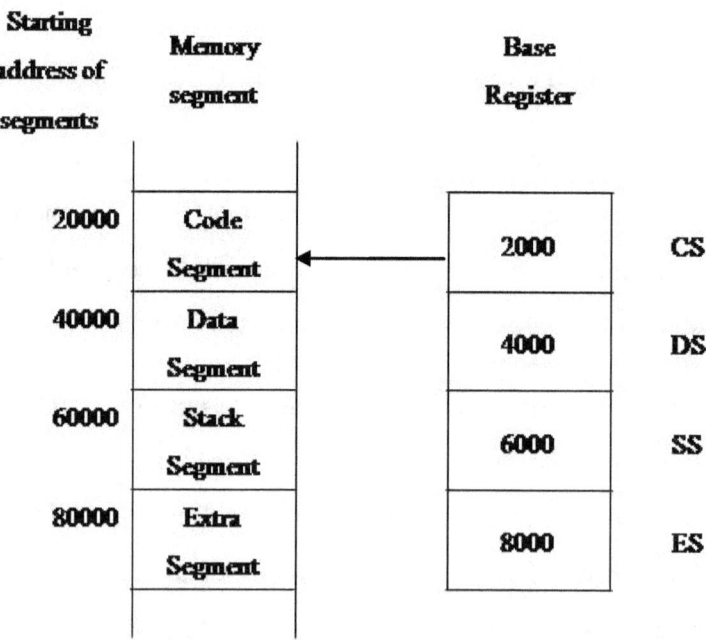

The code segment register is used for address the memory location in the code segment (CS) of the memory where the executable program is stored. The data segment register points to the data segment of the memory where the data resides. The extra segment register also contains data. The stack segment register is used for addressing the memory where the stack data is stored.

4. Instruction pointer (IP)

The instruction pointers contents the offset within the code segment. The instruction pointer acts as a program counter. It indicates the next instruction to be executed. Its contents are automatically incremented when the execution of the program proceeds further.

5. Status flag register

15	14	13	12	11	10	9	8	7	6	5	4	3	2	1	0	Bit no
X	X	X	X	O	D	I	T	S	Z	X	AC	X	P	X	CY	

O	:-	Overflow flag
D	:-	Direction flag
I	:-	Interrupt flag
T	:-	Trap flag
X	:-	Not used
S	:-	Sign flag
Z	:-	Zero flag
AC	:-	Auxiliary carry flag

P   :-   Parity flag

CY   :-   Carry flag

The 8086 has 9 status flags

S   :   This flag is SET when the result of any computation is negative.

Z   :   This flag is SET if the result of computation is zero

P   :   This flag is SET to 1, if the lower byte of the result contains even number of 1.

C   :   This flag is SET when there is carry out of MSB in case of addition or a borrow in case of subtraction.

T   :   This flag is SET when the processor enters the single step execution mode i.e. trap interrupt is generated after execution of each instruction

I   :   This flag is SET when the when the maskable interrupt are recognised by CPU

D   :   This flag is used by string manipulation instructions, when this flag is SET, the string bytes are accused from high memory address to low memory address.

AC :   This flag is SET, if there is carry from the lowest nibble.

O   :   This flag is SET, if an overflow occurs i.e. if the result of a signed operation is large enough to be accommodated in a destination register.

Differences between 8086 & 8088

The architecture of 8086 is similar to 8086 except for two changes :-

1. 8088 has 4-bytes instruction queue

2. 8088 has 8-bit data bus

    The register set of 8088 is exactly the same as 8086 the function of each block is the same as in 8086.

    The 8088 can access only a byte at a time and hence the speed of operation of 8088 reduced as compared to 8086. But 8088 can process the 16-bit data internally. The 8088 has slightly different timing diagram than 8086.

PIN DIAGRAM OF 8086

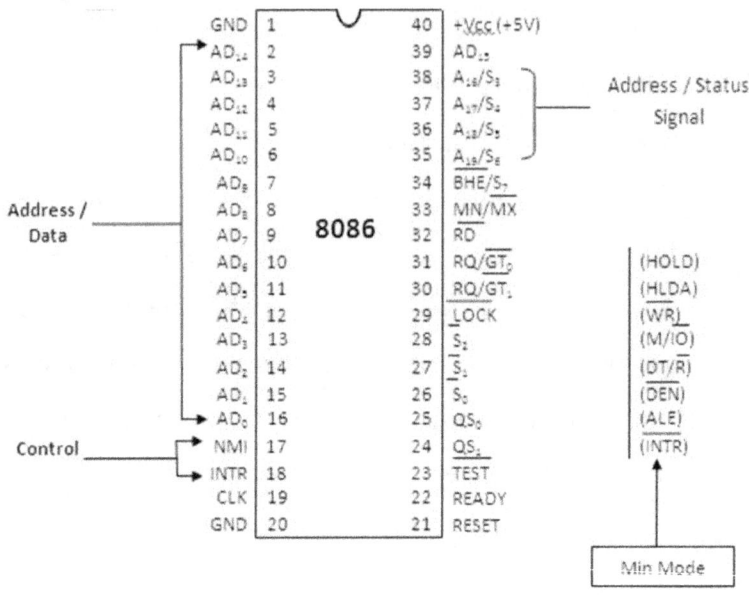

NMI	:	Non Maskable interrupt Input
$\overline{INTR}$	:	Maskable Interrupt Input i.e. request
$\overline{BHE}$	:	Bus High Enable
MN/$\overline{MX}$	:	Min mode / Max mode
$QS_0,$ QS_1	:	Instruction queue status
RQ/$\overline{GT_0}$	:	Request / Grant bus
RQ/$\overline{GT_1}$	:	Request / Grant bus
DT/$\overline{R}$	:	Data Transmit / Receive
$\overline{DEN}$	:	Data Enable

OPERATING MODES OF 8086

There are two modes of operation for Intel 8086, namely: minimum and maximum mode. When only one 8086 CPU is to be used in microcomputer system then 8086 is used in the minimum mode of operation. In this mode the CPU issues the control signals required by memory and I/O devices.

In multiprocessor system 8086 operates in maximum mode. In case of maximum mode of operation, control signals are issued by Intel 8288 bus controller which is used with 8086.

Pin MN/$\overline{MX}$ decides the operating mode of 8086. Pins 24 to 31 have alternate functions for min and max mode.

Pin description for min mode

For minimum mode of operation the pin MN/$\overline{MX}$ is connected to +5v DC power supply. The pins are

$\overline{INTA}$ (O/P) Pin 24 :- Interrupt Acknowlegde

It is active low pin. On receiving interrupt signal the processor issued an Interrupt acknowledge

ALE (O/P) Pin 25 :- Address Latch Enable

It goes high during T_1. The microprocessor sends this signal to latch the address in to Intel 8282/8283 latch.

$\overline{DEN}$ (O/P) Pin 26 :- Data Enable

It is active low pin. When Intel 8286/ 8287 octal bus transreceiver is used then this signal acts as an Output enable signal.

DT/$\overline{R}$ (O/P) Pin 27 :- Data Transmit / Receive

When Intel 8286/ 8287 octal bus transreceiver is used, this signal controls the direction of data flow through the transreceiver. When it is high, data are send out. When it is low, data are received.

M/$\overline{IO}$ (O/P) Pin 28 :- Memory or I/O access

When it is high the CPU wants to access memory. When it is low the CPU wants to access I/O device.

$\overline{WR}$ (O/P) Pin 29 :- Write

When it is low the CPU performs memory or I/O write operation.

HLDA (O/P) Pin 30 :- HOLD acknowledge

It is issued by the processor when it receives HOLD signal. It is active high signal. When HOLD request is removed HLDA goes low

HOLD (I/P) Pin 31 :- HOLD

It is active high signal. When another device in the complex microcomputer system wants to used the address and data bus, it sends a HOLD request through this pin.

Pin description for max mode:

For max mode of operation pin MN/$\overline{MX}$ is made low. It is grounded

Pins are-

Pins 24,25 (QS$_1$, QS$_2$ (O/P)

(Instruction Queue Status)

QS$_1$ QS$_2$

QS$_1$	QS$_2$	
0	0	No Operation
0	1	1-Byte of OP-code from queue
1	0	Empty Queue
1	1	Subsequent byte from queue

Pins 26, 27, 28 ($\bar{S}_0$, $\bar{S}_1$, $\bar{S}_2$(O/P)) (Status Signals)

These signals are connected to the bus controller Intel 8288. The bus controller generates memory and I/O access control signals.

Pin 29 ($\overline{LOCK}$)(O/P)-

It is an active low signal. When it is low then all interrupts are masked and no HOLD request is granted, i.e. in multiprocessor system, all other processors should not ask the CPU for relinquishing bus control.

Pins 30, 31($RQ/\overline{GT}_1$, $RQ/\overline{GT}_0$) (BI-directional)

(Request/ Grant)

(Local bus priority control)

Other processors ask the CPU through these lines to release the local bus. $RQ/\overline{GT}_0$ has higher priority than $RQ/\overline{GT}_1$.

In maximum mode of operation signals $\overline{WR}$, ALE, $\overline{DEN}$, $DT/\overline{R}$, etc. are not available directly from the processor. These signals are available from the controller 8288.

ADDRESSING MODES OF 8086:

Addressing modes indicates a way of locating data or operands . Depending upon the data types used in the instructions and the memory addressing modes, any instruction may belong to one or more addressing modes or some instructions may not belong to any of the addressing mode.

According to the flow of instruction execution, the instructions may be categorized as-

1) Sequential control flow instructions
2) Control transfer instructions

Sequential control flow instruction are the instructions which after execution, transfers control to the next instruction appearing immediately after it in the program, viz. Arithmetic, logical, data transfer and processor control instructions are sequential control flow instructions.

The control transfer instructions transfer the control to some predefined address after their execution. Viz., INT, CALL, RET, JUMP, come under this category.

The addressing modes for sequential control transfer instructions are-

1) **Immediate**

2) **Direct**

3) **Register**

4) **Register Indirect**

5) **Indexed**

6) **Register Relative**

7) **Based Indexed**

8) **Relative based Indexed**

The addressing modes for control transfer instructions are-

1) Inter-segment Direct

2) Inter-segment Indirect

3) Intra-segment Direct

4) Intra-segment Indirect

Explanation

1   Immediate:- In this type of addressing, immediate data is a part of instruction and appears in the form of successive byte or bytes.

    EX: MOV AX, 0005H

0005H is the immediate data. The immediate data may be 8-bit or 16-bit

Instruction

| Data M |

2   Direct:- In direct addressing mode, 16-bit memory address (offset) is directly specified in the instructions as a part of it.

    EX: MOV AX, [5000H]

Here data resides in memory location in data segment (DS), whose effective address may be computed using 5000H as the offset address. The effective address is

EA = 10H * DS + 5000H

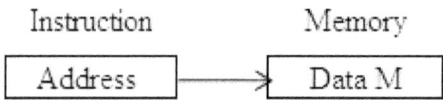

3   Register:- In register addressing mode, the data is stored in a register. All registers except IP may be used in this mode

EX: MOV AX, BX

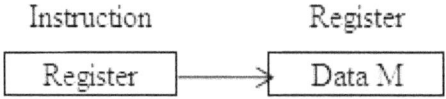

4   Register Indirect:- In this addressing mode, the address of the memory location which contains data or operand is determined in an indirect way, using offset registers. The default segment is either BX or SI or DI register. The default segment is either DS or ES

EX: MOV AX. [BX]

Here, data is present in a memory location in DS whose offset address is in BX.

The effective address is

EA = 10H * DS + [BX]

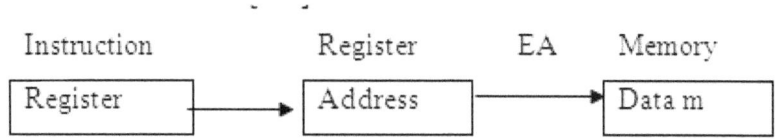

5   Indexed:- In this addressing mode, offset of the operand is stored in one of the index registers. DS & ES are the default segments for index registers SI & DI respectively

EX:MOV AX, [SI]

Here, data is available at an offset address stored in SI in DS.

The effective address is

EA = 10H * DS + [SI]

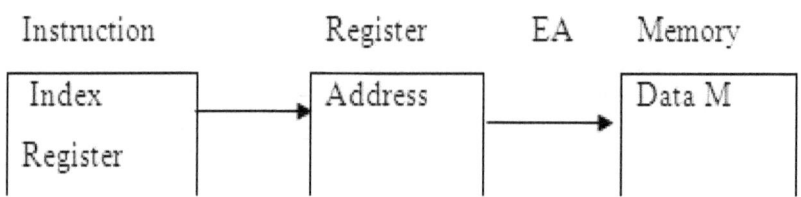

6   Register Relative:- In this addressing mode, the data is available at an effective address formed by adding 8-bit or 16-bit displacement with the content of any one of the registers BX, BP, SI and DI in the default (either DS or ES segments)

EX: MOV AX, 50H [BX]

Effective address is

EA = 10H * DS + 50H + [BX]

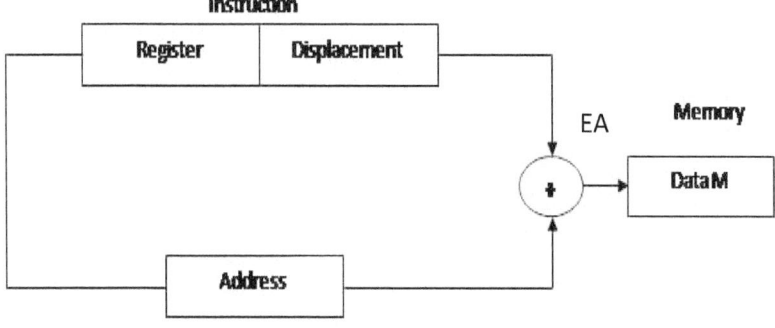

$$EA = \begin{Bmatrix} BX \\ BP \\ SI \\ DI \end{Bmatrix} + \begin{Bmatrix} 8-bit \\ or \\ 16-bit \\ displacement \end{Bmatrix}$$

7 Based Indexed addressing mode:- In this addressing mode, the effective address of data is formed by adding the contents of base register (any one of BX or BP) to the contents of an index registers (any one of SI or DI). The default segment register may be ES or DS.

EX: MOV AX, [BX] [SI]

The effective address is

EA = 10H * DS + [BX] + [SI]

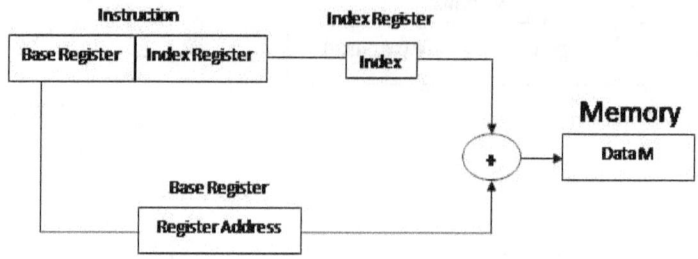

$$EA = \begin{Bmatrix} BX \\ BP \end{Bmatrix} + \begin{Bmatrix} SI \\ DI \end{Bmatrix}$$

8   Relative Based Indexed:- In this addressing mode, the effective address is formed by adding

8-bit or 16-bit displacement with the sum of contents of any one of base registers (BX or BP)

And any one of index registers (SI or DI) in a default segment.

EX: MOV AX, 50H [BX] [SI]

The effective address is

EA = 10H * DS + [BX] + [SI] + 50H

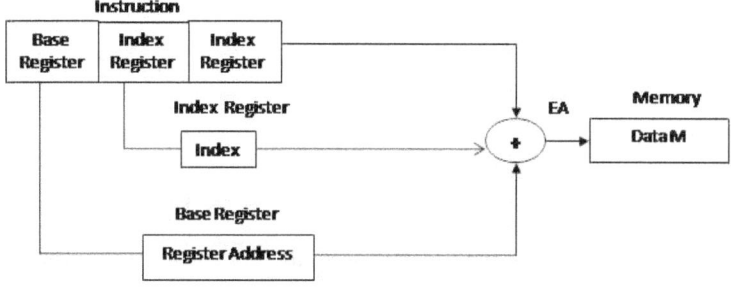

$$EA = \begin{Bmatrix} BX \\ BP \end{Bmatrix} + \begin{Bmatrix} SI \\ DI \end{Bmatrix} \begin{Bmatrix} 8 - bit \\ or \\ 16 - bit \\ displacement \end{Bmatrix}$$

For the control transfer instructions, the addressing modes depends upon whether the destination location is within the same segment or different one. Basically there are two addressing modes for control transfer instructions.

1) Inter-segment

2) Intra-segment

9   Intra-segment Direct mode:-In this mode, the address to which the control is to be transferred lies in the same segment and appears directly in the instruction as an immediate displacement value. In this addressing mode, the

displacement is computed relative to the content of the instruction pointer(IP)

10 Intra-segment indirect mode:-In this mode, the displacement to which the control is to be transferred is in the same segment but t is passed to the instruction indirectly. Here, the branch address is found as the content of a register or memory location. This mode may be used in unconditional branch instruction.

11 Inter-segment direct mode:-In this mode, the address to which the control is to be transferred is in a different segment. This addressing mode provides a means of branching from one code segment to another code segment. Here, the CS & IP of the destination address are specified directly in the instruction.

12 Intersegment indirect mode:-In this mode, the address to which the control is to be transferred lies in different segment and it is passed to the instruction indirectly, i.e., contents of a memory block containing 4 bytes, i.e., IP(LSB), IP(MSB), CS(LSB) and CS(MSB) sequentially.

CHAPTER II

INSTRUCTION SET

INSTRUCTION SET OF 8086

The 8086/8088 instructions are categories into following main types:

1] Data Copy/ Transfer instructions:-

These types of instructions are used to transfer data from source operand to destination operand. All the store, move, load, exchange, input and output instructions belong to this category.

2] Arithmetic & Logical instructions:-

All the instructions performing arithmetic, logical, increment, decrement, compare and scan instructions belong to this category.

3] Branch Instructions:-

These instructions transfer control of execution to the specified address. All the call, jump, interrupt and return instructions belong to this class.

4] Loop Instructions:-

If these instructions have REP prefix with CX used as count register, they can be used to implement unconditional and conditional loops. The LOOP, LOOPNZ an LOOPZ instructions belong to this category. These are useful to implement different loop structures.

5] Machine control instructions:-

These instructions controls the machine status. NOP, HLT, WAIT, & LOCK instructions belong to this class.

6] Flag Manipulation instructions:-

All the instructions which directly affect the flag register, come under this group of instructions. Instructions like CLD, STD, CLI, STI, etc belong to this category of instructions

7] Shift and Rotate instructions:-

These instructions involve the bitwise shifting or rotation in either direction with or without a count in CX.

8] String Instructions:-

These instructions involve various manipulation operations like load, move, can, compare, store etc. These instructions are only to be operated upon the string.

Data Copy/ Transfer instructions:

MOV = Move

PUSH = Push (Memory / Register)

POP = Pop (Register/ Memory)

XCHG = Exchange

IN = Input from

OUT = Output to

XLAT = Translate Byte to AL

LEA = Load EA to register

LDS = Load pointer to DS

LES= Load pointer to ES

LAHF = Load AH with flags

SAHF = Store AH into flags

PUSHF = Push flags

POPF = Pop flags

Arithmetic

ADD = Add

ADC = Add with carry

INC= Increment

AAA = ASCII Adjust for Addition

DAA = Decimal Adjust For Addition

SUB = Subtract

SBB = Subtract with borrow

DEC = Decrement

NEG = Change Sign

CMP = Compare

AAS = ASCII Adjust for subtract

DAS = Decimal Adjust for Subtract

MUL = Multiply (unsigned)

IMUL = Integer multiply (signed)

AAM = ASCII adjust Multiply

DIV = Divide (unsigned)

IDIV = Integer Divide (Signed)

AAD = ASCII Adjust for Divide

CBW = Convert Byte to word

CW = Convert word to Double word

Logical

NOT   =   Invert

SHL/SAL   =   Shift logical/ Arithmetic left

SHR   =   Shift Logical Right

SAR   =   Shift Arithmetic Right

ROL   =   Rotate Left

ROR   =   Rotate Right

RCL   =   Rotate through carry flag left

RCR   =   Rotate through carry right

AND   =   and

TEST   =   AND functions to flags, NO Result

OR   =   or

XOR   =   Exclusive Or

String Manipulations

REP   =   Repeat

MOVS   =   Move Byte/ Word

CMPS   =   Compare (string) Byte/ word

LODS   =   Load byte/wd to AL / AX

STOS   =   Store Byte/ wd from Al / A

Control Transfer

CALL   =   Call

JMP   =   Unconditional Jumps

RET   =   Return from Call

JE/JZ   =   Jump on Equal/ Zero

JL/JNGE   =   Jump on Les/ Not Greater or Equal

JLE/JNG   =   Jump on less or equal/Not greater

JB/JNAE   =   Jump on below/Not above or equal

JBE/JNA   =   Jump on below or equal/Not Above

JO   =   Jump on overflow

JS   =   Jump on sign

JNE/JNZ   =   Jump on not Equal/ Not Zero

JNL/JGE   =   Jump on not les/Greater or Equal

JNLE/JG   =   Jump on Not less or Equal/ Greater

JNB/JAE   =   Jump on Not below/ Above or Equal

JNBE/JA   =   Jump on Not Below or Equal or Above

JNP/JPO   =   Jump on Not par/ Parity odd

JNO   =   Jump on not overflow

JNS   =   Jump on Not Sign

LOOP =   Loop CX times

LOOPZ/LOOPE =   Loop while zero/ Equal

JCXZ =   Jump on CX zero

INT   =   Interrupt

Type specified

INTO =   Interrupt on overflow

IRET =   Interrupt Return

Processor Control

CLC   =   Clear Carry

CMC   =   Complement Carry

STC   =   Set carry

CLD   =   Clear direction

STD   =   Set Direction

CLI   =   Clear interrupt

STI   =   Set Interrupt

HLT   =   Halt

WAIT   =   Wait

ESC   =   Escape (to external Device)

LOCK   =   Bus Lock Prefix

VARIOUS ROTATE INSTRUCTIONS IN 8086 MICROPROCESSOR

ROR :- Rotate Right without carry.

This instruction rotate the contents of the destination operand to the right (bit-wise) either by one or by the count specified in CL, excluding carry.

ROL :- Rotate left without carry

This instruction rotate the contains of the destination operand to the left by the specified count (bit-wise) excluding carry.

RCR :- Rotate Right through carry

This instruction rotates the contents (bit-wise) of the destination operand right by the specified count through carry flag (CF)

RCL:- Rotate Left through carry

This instruction rotates (bit-wise) the contents of the destination operand left by the specified count through the carry flag (CF)

ROR –

Bit position
Operand Count

15	14	13	12	11	10	9	8	7	6	5	4	3	2	1	0	CF
1	0	1													1	X
1	1	0														1

ROL

Bit position
Operand Count

CF	15	14	13	12	11	10	9	8	7	6	5	4	3	2	1	0
X	1	0	1													1
1	1	1	0													1

RCR –

Bit position
Operand Count =1

15	14	13	12	11	10	9	8	7	6	5	4	3	2	1	0	CF
1	0	1													1	0
0	1	0														1

RCL

Bit position
Operand Count =1

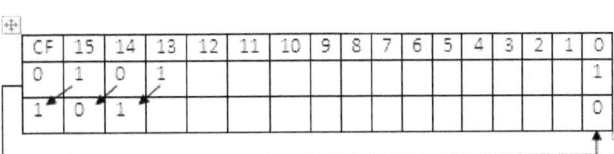

CF	15	14	13	12	11	10	9	8	7	6	5	4	3	2	1	0
0	1	0	1													1
1	0	1														0

DIFFERENT FORMAT OF MOV INSTRUCTION

	MOV Instruction	Effective Address
1)	Immediate MOV AX, 0005H	
2)	Direct MOV AX, [5000H]	EA = 10H * DS + 5000H
3)	Register MOV AX, BX	
4)	Register Indirect MOV AX, [BX]	EA = 10H * DS + [BX]
5)	Indexed MOV AX, [SI]	EA = 10H * DS + [SI]
6)	Register Relative MOV AX, 50H [BX]	EA = 10H * DS + 50H + [BX]
7)	Based Indexed MOV AX, [BX] [SI]	EA = 10H * DS + [BX] + [SI]
8)	Relative based index MOV AX, 50H [BX] [SI]	EA = 10H * DS + [BX] + [SI] + 50H

CONDITIONAL AND UNCONDITIONAL BRACH INSTRUCTION OF 8086

Branch Instructions are of two types –

1) Unconditional control Transfer (Branch) Instructions :-

   In case of unconditional control transfer instruction, the execution control is transferred to the specified location independent of any status or condition. The CS & IP are unconditionally modified to new CS & IP.

   Ex. CALL – This instruction is used to call subroutine from a main program. The address of the procedure subroutine may be specified directly or indirectly depending upon the addressing mode. The mode for them are called as intersegment (in the same segment i.e. near CALL) and intersegment (in another segment i.e. FAR CALL) on execution this instruction stores the incremented IP. (i.e. address of next instructions) & CS onto the stack & load CS & IP register respectively, with the segment & offset addresses of the procedure to be called.

   In case of NEAR CALL it pushes only IP register & in case of FAR CALL it pushes IP & CS both onto the stack.

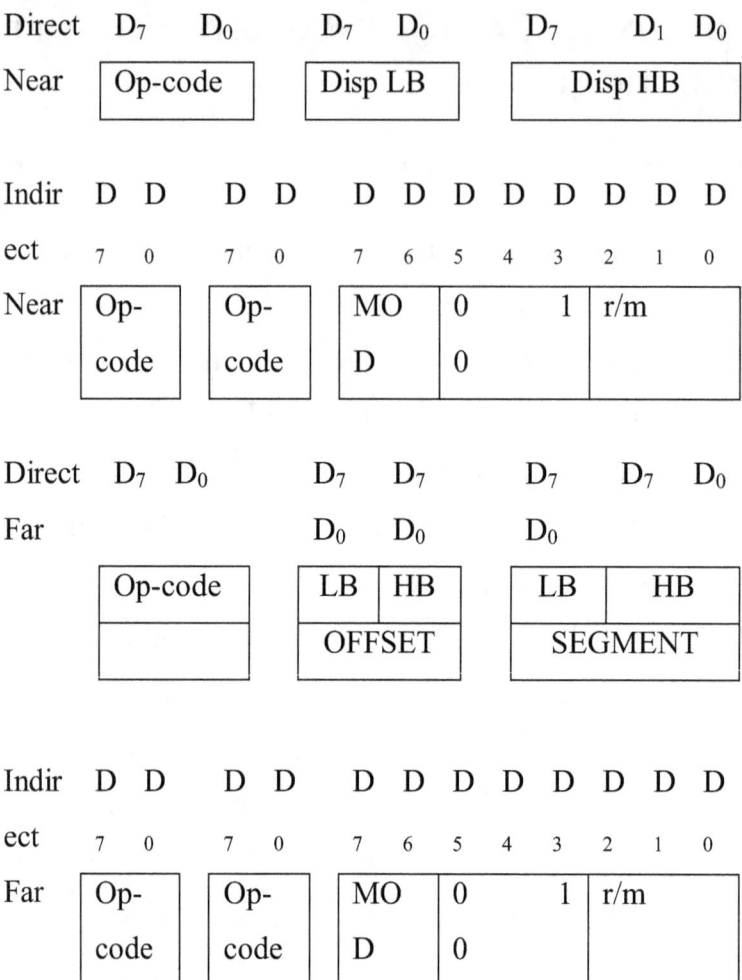

RET :- Return from the procedure

At each CALL instruction, the IP and CS of the next instruction is pushed on to STACK before the control is transferred to the procedure. At the end of the procedure, the RET instruction must be executed. When it is executed, the previously stored contents of IP and CS along with the flags are retrieved into CS, IP and FLAGS

Registers from the STACK and the execution of the main program continues further.

Depending upon the type of procedure and SP contains, the RET instruction is of four types

1. Return within segment
2. Return within segment adding 16-bit immediate displacement to the SP contents
3. Return intersegment
4. Return intersegment adding 16-bit immediate displacement to the SP contents

The other examples of unconditional Branch Instructions are

INT N :- Interrupt Type N

In the interrupt structure of 8086, 256 interrupts are defined corresponding to the types from 00H to FFH. For the execution of this instruction, IF must be enabled.

INT O :- Interrupt On overflow

This command is executed when the overflow flag OF is set

JMP :- Unconditional Jump

This instruction unconditionally transfers the control of execution to the specified address using 8-bit or 16-bit displacement. No flags are affected by this instruction.

LOOP :- Loop unconditionally

This instruction executes the part of the program, the label or address specified in the instruction up to the Loop instruction for CX number of times

CONDITIONAL BRANCH INSTRUCTION

When this instructions are executed, execution control is transferred to the address specified relatively in the instruction provided the condition implicit in the op-code which is specified

JE/JZ   =   Jump on Equal/ Zero

JL/JNGE   =   Jump on Les/ Not Greater or Equal

JLE/JNG   =   Jump on less or equal/Not greater

JB/JNAE   =   Jump on below/Not above or equal

JBE/JNA   =   Jump on below or equal/Not Above

JO   =   Jump on overflow

JS   =   Jump on sign

JNE/JNZ   =   Jump on not Equal/ Not Zero

JNL/JGE   =   Jump on not les/Greater or Equal

JNLE/JG   =   Jump on Not less or Equal/ Greater

JNB/JAE   =   Jump on Not below/ Above or Equal

JNBE/JA   =   Jump on Not Below or Equal or Above

JNP/JPO   =   Jump on Not par/ Parity odd

JNO   =   Jump on not overflow

JNS   =   Jump on Not Sign

FLAG MANIPULATION INSTRUCTIONS

The flag manipulation and processor control instructions control the functioning of the available hardware inside the processor chip

The flag manipulation instructions directly modified some of the flags of 8086

The instructions are

CLC   =   Clear Carry flag

CMC   =   Complement Carry flag

STC   =   Set carry flag

CLD   =   Clear direction flag

STD   =   Set Direction flag

CLI   =   Clear interrupt flag

STI   =   Set Interrupt flag

STRING INSTRUCTIONS

A series of data bytes or words available in memory at consecutive locations to be referred to collectively or individually are called as byte string or word string. The length of a string is usually stored as count in CX register. The incrementing or decrementing of the pointer depends upon the Direction Flag (DF) status.

REP:- Repeat Instruction Prefix

This instruction is used as a prefix to other instructions. The instruction to which the REP prefix is provided is executed repeatedly until the CX register becomes zero. There are two more options of REP instruction

REPE / REPZ :- Repeat operation while Equal / Zero

REPNE/ RENZ :- Repeat operation while Not Equal / Not Zero

MOVSB / MOVSW :- Move String Byte or String Word

This instruction moves a string of bytes / words pointed to by DS : SI pair (source) to the memory location pointed to by ES : DI pair (destination). The REP instruction repeat the trasffer up to the value given in CX.

CMPS :- Compare string byte or string word.

This instruction can be used to compare two strings of bytes or words. The length of the string must be stored in registers CX. The DS : SI & ES : DI points to two strings. The REP instruction repeats the comparison till CX = 0. If both the byte or word strings are equal, zero flag is set.

SCAS : Scan byte or string word.

This instruction scans a string of bytes or words for an operand byte or word specified in AL or AX. The string is pointed to by ES : DI register pair. The length of string is stored in string in CX. The DF (Direction flag) control the mode of scanning. The pointers and counters are updated automatically till the match is found. When the match is found, execution stops and zero flag is set.

LODS :- Load String Byte or String Word

This instruction loads AL / AX register by the contents of string pointed to by DS : SI register pair. The SI is modified automatically depending upon DF. If it is a byte transfer, SI is modified by one and if it is a word transfer, SI is modified by two. No other flags are affected by this instruction.

STOS :- Store string byte or string word

The STOS instruction stores AL/AX register contents to a location in string pointed by ES : DI register pair. The DI is modified accordingly. No flags are affected by this instruction

The SI and DI are modified after each iteration automatically. If DF = 1 then the execution follows auto-decrement mode. If DF = 0 then execution follows auto-increment mode.

MACHINE LANGUAGE

The instructions are hand coded in terms of 0 and 1 (Binary) which are understandable to CPU. This type of language which can understand by the machine is called machine language and the programming procedure is called as machine language programming.

The main advantage of machine language programming is that the memory control is directly in the hands of programmer enabling him to manage the memory of the system more efficiently.

The dis-advantage of machine language programming are

1. The process is complicated and time consuming.
2. The chances of error being committed are more at the machine level in hand coding and entering the program byte – by - byte in to the system.
3. Debugging a program at the machine level is more difficult.
4. The programs are not understood by everyone and results are not stored in a user friendly form.

ASSEMBLY LANGUAGE

The 8086 microprocessor programming is in terms of mnemonics. This language is termed as assembly language. A program called "Assembler " is used to convert mnemonics of instructions along with the data into their equivalent object code modules. These object code modules may further be converted in to executable code using the linker and loader programs. This type of programming is called Assembly Language Programming.

The advantage of assembly language over machine language are

1. Assembly language programming is not so complicated as machine language because the function of coding is performed by a assembler.

2. The chances of error being committed are less because mnemonics are used instead of numerical codes.

3. It is easier to enter an assembly language program.

4. The debugging is easier.

5. Advanced assembler provide facilities like Macros, Lists etc making the task of programming much easier.

6. The memory control is in the hands of user as in machine language

7. The result may be stored in a more user friendly form.

8. The flexibility of programming is more in assembly language

CHAPTER III

ASSEMBER DIRECTIVES

ASSEMBLER DIRECTIVES AND OPERATORS

An assembler is a program used to convert an assembly language program into a equivalent machine code modules which may further be converted into executable codes. It decides the address of each label and substitute the values for each of the constant and variables. It then form the machine code for the mnemonics & data in the assembly language program. For completing all these tasks, an assembler need some hints from the programmer i.e. the required storage for particular constant or variable, logical names of the segments, types of the different routines & modules, end of file etc. These types of hints are given to the assembler using some predefined alphabetical string called assembler directives which help the assembler to correctly understand the ALP & to prepare the codes.

Another type of hints which helps the assembler to assign a particular constant with a label or initialize particular memory location or label with constants is an operator. Operators perform arithmetic & logical task.

DB (Define Byte)

The DB directives is used to reserve byte or bytes of memory location in the available memory. These memory location shows data type which may be constant, variable or string etc.

e.g.   RANKS   DB 01H, 02H, 03H, 04H

         MESSAGE   DB 'GOOD MORNING'

DW (Define Word)

The DW directives serves the same purpose as DB directive but it now makes the assembler reserve the number of memory words (16-bit) instead of bytes.

e.g. WORD   DW 1234H, 4567H, 78ABH, 045CH

DQ (Define Quad word)

This directive is used to direct the assembler to reserve 4 words (8 bytes) of memory for the specified variable and may initialize it with the specified values.

ASSUME (Assume Logical Statement Name)

The ASSUME directive is used to inform the assembler, the name of the logical segments to be assumed for different segments used in the program. In ALP, each segment is given a name viz. for CODE segment CODE and for data segment DATA etc. The ASSUME statement must be at the starting of each ALP.

e.g. ASSUME   CS : CODE

               DS : DATA

It means CS : CODE directs the assembler that the machine codes are available in the segment name CODE & related data is available in the data segment having name DATA.

END (END of program)

The END directive makes the end of ALP It should be the last statement in the file. When assemblers comes across END directive, it ignores the source lines available later on.

ENDP (END of procedure)

In ALP the subroutine are called procedure. The ENDP directive is used to indicate the end of procedure. A procedure may be independent program modules which return particular result or values to the calling programs. A procedure is usually assigned a name is label. To mark the end of a particular procedure, the name of the procedure i.e. label may be appear as a prefix with the directive ENDP e.g. PROCEDURE STAR

.

.

STAR ENDP

END (End of statement)

This directive marks the end of the logical segment. The names appear with the ENDS directive as prefix to mark the end of those particular segments. Any statement appearing after ENDS will be neglected from segment.

e.g. DATA   SEGMENT

.

.

DATA   ENDS

ASSUME   CS : CODE , DS : DATA

CODE   SEGMENT

.

.

CODE   ENDS

END

EVEN (Align on Even memory address)

The EVEN directives updates the location counter to the next even address, if the current location counter contents are not even, and assign the following routine or variable or constant to that address.

e.g. EVEN

PROCEDURE   ROOT

.

.

ROOT   ENDP

EVEN directive checks the contents of location counter and if it is old, it is updated to the next even value.

EQU (Equate)

The directive EQU is used to assign a label with a value or symbol. The use of this directive is just to reduce the recurrence of the numerical values or constants in a program code. The recurring value is assigned with a label & that label is used in place of that numerical value, throughout the program. Using EQU directive, instruction mnemonic can be assigned with a label

e.g. LABEL   EQU 0050H

EXTRN (External) & PUBLIC (Public)

The directive EXTRN inform the assembler that the names, procedures & labels declared after this directive have already been defined in some another ALP module. While the modules where the names, procedure & labels actually appear, they must be declared public using PUBLIC directive.

MODULE 1	SEGMENT
PUBLIC	FACTORIAL FAR
MODULE 1	ENDS
MODULE 2	SEGMENT
EXTRN	FACTORIAL FAR
MODULE 2	ENDS

The program FACTORIAL FAR can be called from MODULE 2 to MODULE 1.

GROUP (Group the related segments)

This directive is used to form logical groups of segments with similar purpose or time. The directive is used to inform the assembler for the logical group. The assembler passes information to the linker / loader to form the code such that the group declared segments or operands must lie within 64 Kbyte memory segment.

PROGRAM GROUP   CODE, DATA , STACK

Now for assume statement we can write

ASSUME   CS : PROGRAM, DS : PROGRAM, SS : PROGRAM

LABEL (Label)

The LABEL directive is used to assign a name to the current content of the location counter. The type of label must be specified i.e. wherather it is NEAR or FAR label, BYTE or WORD label etc.

LENGTH (Byte length of a LABEL)

This directive is not available in MASM. This is used to refer to the length of the data array or a string.

LOCAL :- The label, variables, constants or procedure declared LOCAL in a module are to be used only by that particular module.

LOCAL   a , b, DATA , ARRAY, ROUTINE

NAME (Logical Name of the Module)

The NAME directive is used to assign a name to an ALP module. It may help in the documentation.

OFFSET (Offset of a label)

When the assemblers comes across the OFFSET operator along with a label, it first computes the 16-bit displacement of the particular label & replaces the string 'OFFSET LABEL' by the computed displacement.

ORG (Origin)

The ORG directives direct the assembler to start the memory allotment for the particular segment, block or code form the declared address in ORG statement.

If the ORG statement is not written in the program, the location counter is initialized to 0000. If an ORG 200H statement is present at the starting of the code segment of that module, then the code will start from 200H address in code segment.

PROC (Procedure)

The PROC directive mark the start of a name procedure in the statement. Also, the type NEAR or FAR specify the type of procedure. The NEAR directive is used for the procedure in the same segment of memory while FAR directive is used for the procedure to be called by the program located in the different segment of the memory.

e.g. RESULT   PROC   NEAR

        ROUTINE   PROC   FAR

PTR (Pointer)

The PTR operator is used to declare the type of label, variable or memory operant. The operator PTR is predefined by either BYTE or WORD.

e.g. MOV   AL, BYTE   PTR [SI]

        This specifies move content of memory location address by SI (8-bit) to AL

SEG (Segment of a Label)

The SEG operator is used to decide the segment address of a label, variable or procedure & substitute the segment base address in place of 'SEG' Label.

e.g. MOV   AX,   SEG   ARRAY

    MOV   DS, AX

This statement moves the segment address of ARRAY in which it is appearing, to address AX and then to DS.

SEGMENT (Logical statement)

The SEGMENT directive marks the starting of logical segment. The started segment is also assigned a name i.e. label by this statement.

SHORT

The SHORT operator indicates to assembler that only one byte is required to code the displacement for a jump. This method of specifying the jump address saves the memory. Otherwise the assembler may reserve two bytes for the displacement.

e.g. JMP   SHORT   LABEL

TYPE

The TYPE operator direct the assembler to decide the data of the specified label & replace the TYPE label decided data type. For word type variable, the data type is 2, for double word type it is 4 & for byte type it is 1. Suppose the STRING is a word array. The instruction

      MOV   AX,   TYPE   STRING

moves the value 0002H in AX.

GLOBAL

The label, variable, constant or procedure declared GLOBAL may be used by other module of the program.

ROUTINE   PROC   GLOBAL

'+' & '-' OPARATOR

These operator represent arithmetic addition and subtraction. These are typically used to add or subtract displacement (8 or 16-bit) to base or index registers or stack or base pointers.

e.g. MOV   AL, [SI +2]

MOV   DX, [BX -5]

FAR PTR

This directive indicated the assembler that the label following FAR PTR is not available within the same segment & address of the label is of 32-bit i.e. 2 bytes offset followed by 2 byte segment address

JMP   FAR   PTR   LABEL

CALL   FAR   PTR   ROUNTINE

NEAR PTR

This directive indicates that the label following NEAR PTR is in the same segment & need only 16-bit i.e. 2 byte offset to address it.

e.g. JMP   NEAR   PTR   LABEL

CALL   NEAR   PTR   ROUTINE

MAIN STEPS FOR EXECUTION OF ALP USING MASM

A program called "assembler" is used to convert the mnemonics of instruction along with data & their equivalent object module. These object code modules may further be converted in executable code using the linker & loader programs.

These are two assemblers

1. MASM   :- Microsoft Macro Assembler
2. Turbo Assembler
3. DOS Assembler

MASM is one of the popular assemblers used along with LINK program to structure the code generated by MASM in the form of an executable file. MASM reads source program as it inputs & provides an object file. The LINK accepts the object file produce by MASM as input & produce and EXE file.

While writing / executing an program using assembler the steps are

1. Entering a program
2. Assembling a program
3. Linking a program
4. Debugging a program

1. Entering :- The first step to enter a data use text editor e.g. Norton Editor (NE), TURBO-C (TC), EDLIN etc. For every assembly language program, the execution .ASM must be there. The command line for this is

C>   NE KMB.ASM

KMB is a file name. store this file with command F3-E. This will generate new copy of the program in secondary storage. Then quit the NE with command NE-Q

2. Assembling :- The main task of any assembler program is to accept the text – assembly language program file as an input and prepare an object file. The object file is created with the entered name & the .OBJ extension. The .OBJ file contain the coded object module of the program to be assembled. The listing file is automatically generated in the assembly process. The listing file is identified by the entered or the source file name & an extension .LST. This file contains the total offset map of the source file including labels, offset address, op-code memory allotment for labels & directives & relocation information. The cross-reference file name (.CRF) is also entered in the same way as for listing file. This file is used for debugging the source program. It contain the statistical information size of the file in bytes, number of labels, list of labels, routines to be called etc. about the source program. Afterward these file are given to the linker.

```
C> MASM

Source filename          [ .ASM]:
Object filename          [ FILE .OBJ]:
List filename            [ NUL .LIST ] :
List filename            [ NUL .CRF] :
```

3. Linking :- The DOS linking program LINK.EXE links the different module of a source program & function library routines to generate an integrated executable code of the source program.

```
C> LINK

Object Module            [ .OBJ ] :
Run File                 [ .EXE ] :
List File                [ NUL .MAP ]:
Libraries                [ LIB ] :
```

4. Debug :-

DEBUG.COM is a DOS utility that felicitates the debugging & trouble shooting of ALP. The DEBUG utility enables to control the processors resources & memory resource management functions.

The DEBUG command at DOS prompt is as follows –

```
C> DEBUG
              -     <ENTER>
             -R   <ENTER>
```

'-' (dash) display signals the successful operation of DEBUG prompt for debugging commands.

A valid command is accepted using <ENTER> key. The DEBUG program may be used either to debug a source program or to observe the result of execution of an .EXE file with the help of .LST file & the commands. The .LST file shows the offset address allotments for the result variables of a program in particular segment. After execution of the program, the offset address of the result variable may be observed using D command. The result available in registers may be observed using R command.

Command Character	Format	Function
-R	<ENTER>	Display all Registers & Flags
-R	reg <ENTER>	Display specified register contents & modify with the entered new content
-D	<ENTER>	Display 128 memory location of RAM staring from current display pointer
-D	SEG : OFFSET 1 OFFSET 2 <ENTER>	Display memory contents in SEG from OFFSET1 to OFFSET2

-E	<ENTER>	Enter Hexa data at current display pointer SEG : OFFSET
-E	SEG : OFFSET1 <ENTER>	Enter data at SEG : OFFSET1 byte by byte
-a	<ENTER>	Assemble from the current CS : IP
-a	SEG : OFFSET <ENTER>	Assemble the entered instruction from SEG : OFFSET address
-u	<ENTER>	Unassembled from the current CS : IP
-u	SEG : OFFSET <ENTER>	Unassembled from the address SEG : OFFSET
-g	=OFFSET <ENTER>	Execute from OFFSET in the current CS
-q	<ENTER>	Quit the DEBUG & return to DOS
-T	SEG : OFFSET <ENTER>	Trace the program execution by single stepping starting from the address SEG : OFFSET
-m	SEG : OFFSET1 OFFSET2 NB <ENTER>	Move NB byte from OFFSET2 in segment SEG
-C	SEG : OFFSET1 OFFSET2 NB <ENTER>	Copy NB from OFFSET1 to OFFSET2 in segment SEG

CHAPTER IV

STACK

STACK:

The stack is a block of memory that may be used for temporarily storing the content s of register inside the CPU. The stack is a block of memory locations which is accessed by using the SP and SS register. As we go on storing the data words onto the stack, the pointer goes on incrementing as we go on retrieving the word data. The process of storing the data in the stack is called "pushing into" the stack and the reserve process of transferring the data back from the stack to the CPU register is known as "popping of" the stack. The stack is last-in first-out(LIFO) data segment.

After executing the main program up to the call instruction, the control will be transferred to the subroutine address. Now, the microprocessor must know where the control is to be returned, after the execution of the subroutine. A similar problem may arises while handling interrupts .This address of re-entry into the main program may be stored on to the stack. Also, the stack is useful for storing the register status of the processor at the time of calling a subroutine and getting it back at the time of returning, so that the register s or memory locations already used during the main program can be, reused by the subroutine without any loss of data .The stack provides a temporary storage of data in these cases.

The stack pointer(SP) is 16-bit register that contains the offset of address that lies in the stack segment. The stack segment is a memory block of maximum 64kbytes locations. The stack segment register(SS) contains the base address of the stack segment in the memory. The stack segment register(SS) ant the stack pointer register (SP) together address the stack-top.

LET

SS:-5000H

SP:-2050H

SS :-   0101   0000   0000   0000

10H*SS:- 0101   0000   0000   0000   0000

+

SP:-   0101   0000   0101   0000

Stack-top 0101   0010   0000   0101   0000

Address   5   2   0   5   0

Thus, stack top address is 52020H.

STACK STRUCTURE OF 8086 :-

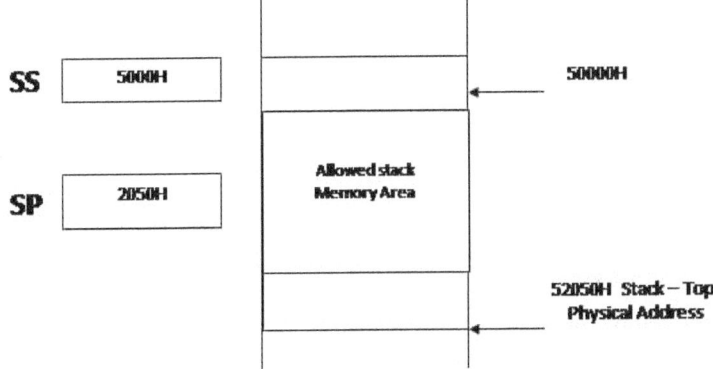

If the stack top points to memory location 52020H it means previously pushed data is available at 52020H.The next 16-bit push operation will decrement the stack pointer will decrement by two, so that it will point to the new stack top 5204EH, and the decremented contents of SP will be 204EH.This location will be now be occupied by recently pushed data.

Thus for a selected value SS the maximum value of SP=FFFFH and the segment can have maximum 64K locations.

Effect of PUSH & POP on SP

For PUSH AX & POP BX

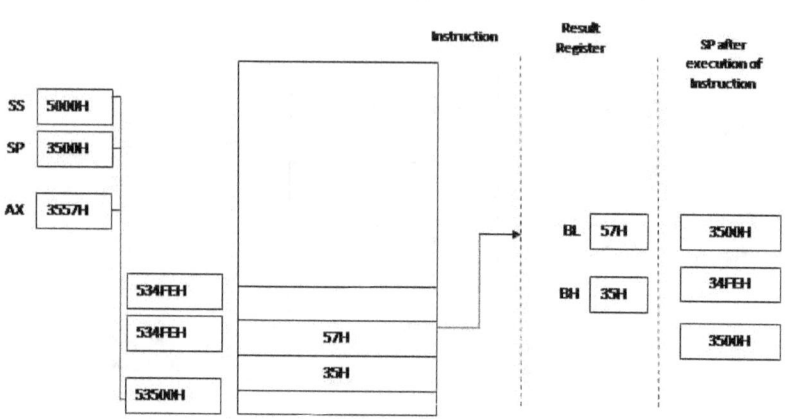

At the start of the subroutine, all the registers contents of the main program may be pushed onto the stack one by one. Thus all the register can copied to the stack. Now these registers may be used by the subroutine, since their original contents are saved onto the stack. At the end of the execution of the subroutine, all the registers can get back their original contents by popping the data from the stack. The sequence of popping is exactly the reverse of the pushing sequence.

USE OF STACK IN PROGRAMMING

In a program the stack can be defined in a similar way as data segment. The assume directive directs the name of the stack segment to the assembler. The SS register must be initialized in the program.

After a subroutine is called using the call instruction, the IP is incremented to the next instruction. Then the contents of IP, CS and flag registers are pushed automatically to the stack. The control is then transferred to the specified address in the call instruction.ie the starting address of the subroutine. Then the subroutine is executed. These should be an equal numbers of push and pop instructions in the subroutine that has to be executed so that the sp contents at the time of calling the subroutine must be equal to the contents of sp at the time of executing the ret instruction, at the end of the subroutine.

INTERRUPTS AND INTERRUPT SERVICE ROUTINES (ISR)-

While the CPU is executing a program an 'interrupt' breaks the normal sequence of execution of instructions, diverts its execution to some other program called interrupt service Routine(ISR).after executing ISR, the control is transferred back again to the main program which was being executed at the time of interruption.

Whenever, a number of devices interrupt a CPU at a time, if a processor is able to handle them properly, it is said to have multiple interrupt processing capability.

There are two type of interrupts

1): External Interrupt

2): Internal Interrupt

In external interrupt, an external device or a signal interrupts the processor from outside viz keyboard interrupt, NMI pin. The internal interrupt is generated internally by the processor circuit or by the execution of interrupt instruction viz divide by zero interrupt, overflow interrupt, interrupt due to INT instructions TRAP etc.

In 8086 microprocessor there are two interrupt pins-NMI(non-maskable interrupt) and INTR(interrupt request).

NMI has the highest priority among the external interrupt.ie NMI input cannot be masked or disable by any means.

INTR has lower priority as compared to NMI. The INTR interrupt may be masked using interrupt flag(IF).The INTR is of 256 types. The INTR types may be from 00H to FFH(0 to 255).If more than one type of INTR interrupt occurs at a time, then an external chip called programmable interrupt controller is required to handle them. IF is SET, the processor is ready to respond to any INTR interrupt .if the IF is reset, the processor will not serve any interrupt appearing at this pin. However, once the processor responds to an INTR signal, the IF is automatically RESET .

TRAP(single set -type1) is an internal interrupt having the highest priority amongst all interrupt except divide by zero(Type 0) exception.

INTERRUPT CYCLE OF 8086:-

Suppose an external device interrupts the CPU at the interrupt pin, either NMI or INTR of 8086, while the CPU is executing an instruction of a program. The CPU first completes the execution of the current instruction .The IP is then incremented to point to the next instruction .The CPU then acknowledge the requesting device on its INTA pin immediately if it is a NMI,TRAP or divide by zero interrupt. If it is INT request, the CPU checks the IF flag. If the IF is set, the interrupt request is acknowledge using INTA pin. If the IF is not set, the interrupt requests are ignored. The response to the NMI,TRAP and divide by zero interrupt requests are independent of IF flag.

After an interrupt acknowledged the CPU computes the vector address from the type of interrupt that may be passed to the interrupt structure of the CPU internally (software interrupts, NMI, TRAP, divide by zero interrupt) or externally i.e. from an interrupts controller in case of external interrupts. The content of IP and CS are next pushed to the stack. The contents of IP and CS now point to the address of the next instruction of the main program from which the execution is to be continued after executing the ISR. The PSW is also pushed to the stack. The interrupt flag (IF) is cleared. The IF is also cleared, after every response to the single step interrupt. The control is then transferred to the interrupt service routine for showing the interrupting device.

At the end of ISR the last instruction should be IRET. When the CPU executes IRET, the contents of flags, IP and CS which were saved at the start by the call instruction are now retrieved to the respective registers. The execution continues onwards from this address, received by IP and CS.

Every external and internal interrupt is assigned with a type (N). Intel has reserved 1024 location for storing the interrupt vector table. The 8086 support a total of 256 types of interrupt i.e. from 00H to FFH. Each interrupt service requires 4 bytes i.e. 2 bytes each for IP and CS of its ISR. Thus a total of 1024 bytes are required for 256 interrupt types, hence the interrupt vector table starts at location 0000:0000 and ends at 0000:03FFFH. The execution automatically starts from the new CS : IP.

INTERRUPT PROGRAMMING:-

While programming for any type of interrupt, the programmer must set the interrupt vector table with proper CS and IP addresses of the interrupt service routine. The method of defining the interrupt service routine for software as well as hardware interrupt is same. It is assume that the interrupt vector table is initialized suitably to point to the interrupt service routine.

The execution sequence of software interrupt is as follows-

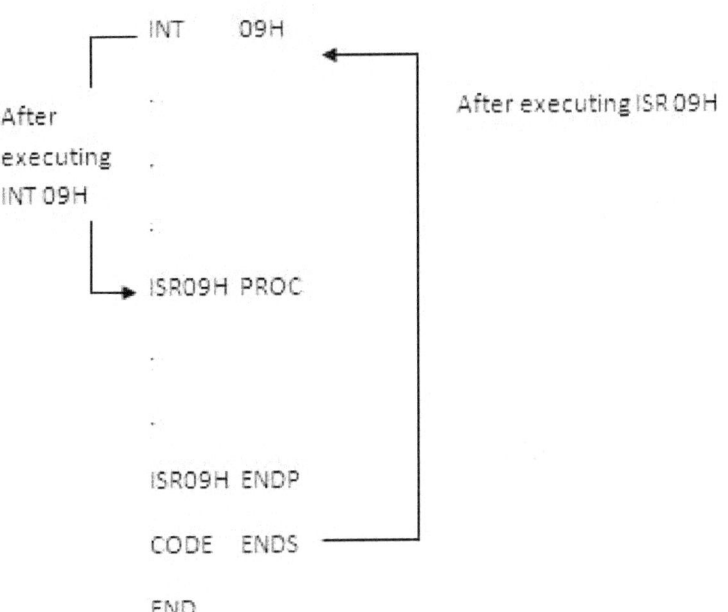

Assume CS : CODE, DS : DATA

DATA   SEGMENT

.

DATA   ENDS

CODE   SEGMENT

INT   09H

After executing INT 09H

After executing ISR 09H

ISR09H PROC

.

ISR09H ENDP

CODE   ENDS

END

MACROS:-

The main program calls subroutine and interrupt service routine and may refers the result of their execution for further processing. The subroutine and interrupt service routine are assigned labels for reference.

The macro is a similar concept. Suppose, a number of instruction are repeating through the main program then the listing becomes lengthy. So a macro definition i.e. label is assigned with the repeatedly appearing string of instruction. The process of assigning a label or macro name to the string is called defining a macro. A macro within a macro is called a nested macro. The macro name or macro definition is then used throughout the main program to refer to that string of instructions.

Defining a macro:-

A macro can be defined anywhere in the program using the directive macro and ENDM. The syntax is as given-

```
DISPLAY     MACRO

            MOV   AX,  SEG MSG

            MOV   DS,  AX

            MOV   DX,  OFFSET MSG

            MOV   AH,  09H

            INT   21H

ENDM
```

A macro may be used in the data segment also. A macro may also be used to represent statements and directives. The concept of macro remains the same, independent of its contains.

There may be more than one parameter to be passed to the macro and each of them is liable to be changed.

EX: for macro containing string

STRINGS MACRO

MSG1 DB 0AH, 0DH, "PROGRAM TERMINATED NORMALLY",0AH, 0DH,"S"

MSG2 DB 0AH, 0DH, "RETRY ABOUT FAIL",0AH,0DH,"S"

ENDM

COMPARISION BETWEEN MACROS AND SUBROUTINE

MACRO	SUBROUTINE
1) The complete code of the instruction string is inserted at each place where the macro-name appears. Hence EXE file becomes lengthy.	1) In case of subroutine the executable code becomes smaller as the subroutine appears only once in a complete code. The EXE file is smaller as compare to macro.
2) Macro does not utilise the service of stack	2) Subroutine utilises the stack service
3) There is no question of transfer of control as the program using macro inserts the complete code of macro at every reference of macro-name	3) The control is transferred to a subroutine whenever it is called and this utilises the stack service
4) The macro requires more memory space	4) The programme using subroutine requires less memory space for execution
5) Macro requires less time for execution	5) Subroutine requires more time for execution
6) It does not contain CALL and RET instruction.	6) It contains CALL and RET instruction.

TIMING AND DELAY

Every instruction requires a definite numbers of clock cycles for its execution. Thus every instruction requires a fixed amount of time ie multiplication of number of clock cycles required for the execution of the instruction and the period of the clock at which the microprocessor is running. The duration required for the execution of instruction can be used to derive required delays. A sequence of instruction, if executed by a microprocessor, will require a time duration that is the sum of all individuals time duration required for execution of each instruction.

In a loop program, the number of instruction in the program may be less but the number of instruction in the program may be less but the number of instruction actually executed by microprocessor depend on the loop count. Also in case of subroutine and interrupt service routines the actual number of instruction executed by the microprocessor depends on the procedure or interrupt service routine length along with main calling program.

The procedure of generating delay using a microprocessor based system is

1) Determine the exact required delay.

2) Select instruction for delay loop. The main program must not be modified by the delay routine.

3) Find out the number of clock states required for execution of each of the selected delay loop instructions.

4) Find out the period of the clock frequency at which microprocessor is running is duration of a clock state (T).

5) Find out the time required for the execution of the loop once by multiplying the period T with the number of clock states required (n) to execute the delay loop once.

6) Find out the count (N) by dividing the required time delay Td by the duration for execution of the loop once (n*T).

$$\text{Count } N = \text{required delay}(Td)/n*T$$

CHAPTER V

ASSEMBLY LANGUAGE PROGRAMMING (ALP)

Assembly Language Program:

1.Write a program in assembly language to find the addition of two 8-bit numbers.

DATA	SEGMENT
	A DB 5H
	B DB 4H
	SUM DB (?)
DATA	ENDS
CODE	SEGMENT
ASSUME	CS:CODE, DS : DATA

START: MOV AX, DATA ;

MOV DS, AX ;

MOV AL, A ;

MOV BL, B ;

ADD AL, BL ;

MOV SUM, AL ;

MOV AX, 4C00H ;

INT 21H ;

CODE ENDS

END START

2.Write a program in assembly language to find the subtraction of 8-bit number.

DATA SEGMENT

A DB 6H

B DB 7H

RESULT DB (?)

DATA ENDS

ASSUME CS : CODE , DS : DATA

```
START :    MOV AX, DATA  ;

           MOV DS, AX;

           MOV AL, A;

           MOV BL, B;

           SUB AL, BL ;

           MOV AH, 4CH ;

           INT 21H ;

CODE       ENDS

           END    START
```

3.Write a program in assembly language to find the addition of 16-bit numbers.

```
DATA   SEGMENT

       A DW 5H

       B DW 4H

       SUM  DW(?)

DATA   ENDS

CODE   SEGMENT

ASSUME    CS:CODE,  DS:DATA
```

```
START :     MOV AX, DATA ;

            MOV DS, AX ;

            MOV AX, A ;

            MOV BX, B ;

            ADD AX, BX ;

            MOV SUM, AX ;

            MOV AX, 4C00H ;

            INT 21H ;

CODE        ENDS

            END  START
```

4.Write an assembly language program of 8086 to perform addition, subtraction, multiplication and division of given operands.

```
DATA        SEGMENT

OPR 1       EQU 20H

OPR 2       EQU 2H

SUM         DW 01 DUP (00)

SUBT        DW 01 DUP (00)
```

PROD   DW 01 DUP (00)

DIVS   DW 01 DUP (00)

DATA ENDS

CODE SEGMENT

ASSUME   CS:CODE   DS:DATA

START: MOV   AX, DATA

    MOV   DS, AX

    MOV   AL, OPR 1

    MOV   BL, OPR 2

    ADD   AL, BL

    DAA

    MOV BYTE PTR SUM, AL

    JNC MSBO

    INC[SUM + 1]

MSBO : XOR   AL, AL

    MOV AL, OPR 1

    SUB AL, BL

    DAS

```
        MOV BYTE PTR     SUBT , AL

        JNB MSBI

        INC [SUBT + 1]

MSB1 : XOR      AL,AL

        MOV     AL,OPR 1

        MUL     BL

        MOV WORD PTR PROD, AX

        XOR     AH, AH

        MOV     AL, OPR 1

        DIV BL

        MOV BYTE PTR DIVS, AL

        MOV AH , 4CH

        INT 21H

CODE   ENDS

END    START
```

5. Write an assembly language program of 8086 to perform a 1 byte BCD addition.

DATA   SEGMENT

OPR 1   EQU 59H

OPR 2   EQU 35H

RESULT   DB 02   DUP(00)

DATA   ENDS

CODE SEGMENT

ASSUME   CS:CODE , DS : DATA

START: MOV   AX, DATA

    MOV   DS, AX

    MOV   BL, OPR 1

    XOR   AL, AL

    MOV   AL, OPR 2

    ADD   AL, BL

    DAA

    MOV   RESULT   AL

    JNCMSBO

    INC[RESULT + 1]

MSBO : MOV   AH, 4CH

   INT 21H

CODE   ENDS

END   START

6.Program to determine average of 2no's

ABSTRACT – This program procedure is used for average of 2 no's

Register used – AX, BH, BL

DATA	SEGMENT				
HI -	NUM	DB	92H		
LO-	NUM	DB	52H		
AVG-	NUM	DB	?		
DATA	ENDS				
ASSUME	CS : CODE	DS:			
		DATA			
START :	MOV	AX,	DATA ;	Initializing data segment register	
	MOV	DS,	AX;		
	MOV	AL,	HI- NUM;	First no in Al register	
	ADD	AL,	LO- NUM;	Addition of second no	
	MOV	AH,	00H;	Clear AH register	
	ADC	AH,	00H;	Add carry to AH register	
	MOV	BL,	02H;	Store 02H in BL	

DIV	BL	;	Divide AX by BL
MOV	[AVG-NUM], AL		Store the result
	;		

CODE   ENDS

END   START

7.Program to find smallest of 2 number

ABSTRACT – This program procedure for smallest of two number

Register used – AX, BL, DS

DATA	SEGMENT		
	NUM1	DB	92H
	NUM2	DB	52H
	SMALL-NUM	DB	?
DATA	ENDS		
CODE	SEGMENT		
ASSUME	CS : CODE	DS:	
		DATA	

START :	MOV	AX,	DATA ;	Initializing data
	MOV	DS,	AX;	segment register
	MOV	AL,	NUM1;	First no in Al register
	MOV	BL,	NUM2;	Second no in BL
	COMP	AL,	BL;	Compare two no's
	JC	NEXT	;	Jump on next step if carry is generated

	MOV	AL,	BL;	Transfer AL to BL
NEXT :	MOV	[SMALL-NUM],		Store the Small
		AL		no.
CODE	ENDS			
END	START			

8.Program to find largest of 2 number

ABSTRACT – This program procedure for largest of two number
Register used – AX, BL, DS

DATA	SEGMENT			
	NUM1	DB	92H	
	NUM2	DB	52H	
	LARGE-NUM	DB	?	
DATA	ENDS			
CODE	SEGMENT			
ASSUME	CS : CODE	DS:		
		DATA		

START :	MOV	AX,	DATA ;⎫	Initializing data
	MOV	DS,	AX; ⎭	segment register
	MOV	AL,	NUM1;	First no in Al register
	MOV	BL,	NUM2;	Second no in BL
	COMP	AL,	BL;	Compare two no's
	JNC	NEXT	;	Jump on next step

			if carry is generated
	MOV	AL, BL;	Transfer AL to BL
NEXT :	MOV	[LARGE-NUM], AL	Store the Small no.
CODE	ENDS		
END	START		

9. **Write a program to ADD series of prices in memory & store result in a memory NET.**

ABSTRACT – This program procedure is used for addition of series of prices.

Register used – AX, CL, DL, BX

ARRAY		
SEGMENT ;		
PRICES	DB	21H, 42H, 36H, 28H, 15H, 35H, 22H
NET	DW	?
ARRAY	ENDS	
CODE	SEGMENT	
ASSUME	CS : CODE	DS: ARRAY
START :	MOV	AX, ARRAY ; } Initializing data segment register

	MOV	DS,	AX;	
	MOV	AX,	0000H;	Clear AX register
	LEA	BX,	PRICE;	Load effective address to register BX
	MOV	CL	07H;	Store 07 count to CL register
NEXT :	MOV	DL,	[BX];	Move content of BX to DL
	ADD	AL,	DL;	Add AL with DL
	DAA		;	Decimal adjust after addition
	JNC	NEXT 1 ;		Jump if not carry to next
	INC	AH ;		Increment AH
NEXT 1 :	MOV	AX;		Move content of AX to NET
	[NET]			
	INC	BX;		Increment BX
	DCR	CL;		Decrement CL
	JNZ	NEXT;		Jump if not zero to next
CODE	ENDS			
END	START			

10. Factorial of an 8 bit number. Using procedure. Number read from keyboard. ; Result stored in memory.

MODEL SMALL

DISPLAY MACRO MSG

CLD ;

    MOV   DX,OFFSET MSG ;

    MOV   AH,09H ;

    INT   21H ;

ENDM

DATA   SEGMENT

MSG1   DB 10,13," ENTER A NUMBER : $" ;

MSG2   DB 10,13," OUT OF RANGE $" ;

NUM   DB ? ;

RESULT DW 02H DUP(?) ;

DATA   ENDS

CODE SEGMENT

```
START :   MOV          AX, DATA ;

          MOV          DS , AX ;

          DISPLAY          MSG1 ;

          MOV          AH,01H ;

          INT          21H ;

          MOV          NUM , AL ;

          CALL          ASCII_TO_HEX ;

          CMP          AL , 0CH ;

          JA           L1 ;

          CALL          FACTORIAL ;

          JMP          L2 ;

ASCII_TO_HEX PROC NEAR ;

          MOV          AL,NUM ;

          CMP          AL,39H ;

          JA        S1;

          SUB          AL,30H ;
```

```
        JMP             S2 ;

S1:     SUB             AL,37H ;

S2:     MOV             NUM,AL ;

        RET

FACTORIAL PROC NEAR

        XOR             AX,AX ;

        XOR             DX,DX ;

        XOR             BX,BX ;

        XOR             CX,CX ;

        MOV             AL,NUM;

        MOV             BL,AL;

        CMP             AL,01H;

        JBE             NEXT2;

        DEC             BX;

UP:     MOV             CX,AX;

        MOV             AX,DX;
```

```
        MUL             BX;

        XCHG            AX,CX;

        MUL             BX;

        ADD             DX,CX;

        DEC             BX;

        JNZ UP;

        MOV             RESULT,AX;

        MOV             RESULT+2,DX;

        JMP             LAST;

NEXT2:  MOV             RESULT,0001H;

        MOV             RESULT+2,0000H;

LAST:   RET;

L1: DISPLAY    MSG2;

L2: MOV        AH,04CH;

        INT         21H;

        END
```

11. LCM and GCD of two 16 bit numbers. Display results on screen. Numbers are stored in memory.

MODEL SMALL

DISPLAY MACRO MSG

      MOV   DX,OFFSET MSG;

      MOV   AH,09H;

      INT   21H;

ENDM

      DATA   SEGMENT

      NUM1   DW ?;

      NUM2   DW ?;

      LCM   DW 02H DUP(?);

      GCD   DW ?;

      MSG1   DB 10,13," THE LCM OF THE TWO NUMBERS IS : $";

      MSG2   DB 10,13," THE GCD OF THE TWO NUMBERS IS : $";

```
LCM_DISPLAY         DB 08H DUP(?) DB '$' ;

GCD_DISPLAY         DB 04H DUP(?) DB '$' ;

DATA                ENDS

CODE                SEGMENT

    MOV             AX, DATA ;

    MOV             DS,AX ;

    XOR             AX,AX ;

CALL LCM_PROCEDURE;

CALL GCD_PROCEDURE ;

    LEA             SI,LCM+3 ;

        LEA         DI,LCM_DISPLAY ;

        MOV         CH,04H ;

BACK:   MOV         AL,BYTE PTR[SI] ;

        CALL        HEX_TO_ASCII ;

        MOV         BYTE PTR[DI],AH ;

        INC         DI ;
```

```
MOV        BYTE PTR[DI],AL ;

DEC        SI ;

INC        DI ;

DEC        CH ;

JNZ        BACK ;

MOV        AL,GCD+1 ;

CALL       HEX_TO_ASCII ;

MOV        GCD_DISPLAY,AH ;

MOV        GCD_DISPLAY+1,AL ;

MOV        AL,GCD ;

CALL       HEX_TO_ASCII ;

MOV        GCD_DISPLAY+2,AH ;

MOV        GCD_DISPLAY+3,AL ;

DISPLAY    MSG1 ;

DISPLAY    LCM_DISPLAY ;

DISPLAY    MSG2 ;
```

```
        DISPLAY        GCD_DISPLAY ;

        JMP            EXIT ;

        LCM_PROCEDURE PROC NEAR

        XOR            DX,DX ;

        MOV            AX,NUM1 ;

        MOV            BX,NUM2 ;

        CMP            AX,BX ;

        JAE            L1 ;

        XCHG           AX,BX ;

L1:     MOV            CX,AX ;

UP:     MOV            LCM,AX ;

        MOV             LCM+2,DX ;

        DIV            BX ;

        CMP            DX,0000H ;

        JZ             L2 ;

        MOV            AX,LCM ;
```

```
        MOV        DX,LCM+2 ;

        ADD        AX,CX ;

        ADC        DX,0000H ;

        JMP        UP ;

L2:     RET

        GCD_PROCEDURE PROC NEAR

        XOR        AX,AX ;

        XOR        BX,BX ;

        XOR        CX,CX ;

        XOR        DX,DX ;

        MOV        AX,NUM1 ;

        MOV        BX,NUM2 ;

        CMP        AX,BX ;

        JAE        G1 ;

        XCHG       AX,BX ;

G1:     XOR        DX,DX ;
```

```
        MOV         CX,AX ;

        DIV         BX ;

        CMP         DX,0000H ;

        JZ          G2 ;

        MOV         AX,BX ;

        MOV         BX,DX ;

        JMP         G1 ;

G2:     MOV         GCD,BX ;

        RET

        HEX_TO_ASCII PROC NEAR

        MOV         AH,AL ;

        MOV         CL,04H ;

        ROR         AH,CL ;

        AND         AH,0FH ;

        CMP         AH,09H ;

        JBE         H1 ;
```

```
            ADD         AH,37H ;

            JMP         H2 ;

H1:         ADD         AH,30H ;

H2:         AND         AL,0FH ;

            CMP         AL, 09H ;

            JBE         H3 ;

            ADD         AL,37H ;

            JMP         H4 ;

H3:         ADD         AL,30H ;

H4:         RET

EXIT:       MOV         AH,04CH ;

            INT         21H ;

            END
```

12. Write a code fragment to read a character from the keyboard:

MOV   AH, 1H ; keyboard input subprogram

INT   21H ; character input

; character is stored in al

13   Reading and displaying a character:

MOV   AH, 1H ; keyboard input subprogram

INT   21H ; read character INTo al

MOV   DL, AL ; copy character to dl

MOV   AH, 2H ; character output subprogram

INT   21H ; display character in dl

14. Generation of first N prime numbers.

MODEL SMALL

DATA   SEGMENT

```
ARRAY          DB 01H  DB 02H DB 50H

DUP(?)N    DB 0FFH

CODE   SEGMENT

START : MOV      AX,DATA

        MOV      DS,AX

        XOR      AX,AX

        LEA      SI,ARRAY+2

        MOV      CX,0003H

 TOP:   MOV      BL,02H

UP:     MOV      AX,CX

        DIV   BL

        CMP   AH,00H

        JZ    INVALID

        INC   BL

        CMP      BL,CL

        JB    UP
```

```
        MOV     BYTE PTR[SI],CL

        INC     SI

        INC     DL ; TO STORE THE COUNT

INVALID:INC     CL

        CMP     DL,N

        JNZ     TOP

        MOV     AH,04CH

        INT     21H

        END
```

15. **8086 program to find out whether the given year is a leap year or not.**

```
        ASSUME DS:DATA1,CS:CODE1
        DATA1 SEGMENT
            MSG     DB 0AH,0DH,'ENTER THE YEAR$'
            NUMBER  DB 6,0,6 DUP('$')
            YS      DB 0AH,0DH,'YES,IT IS A LEAP YEAR$'
            N       DB 0AH,0DH,'NO,IT IS NOT A LEAP
        YEAR$'
```

DATA1 ENDS

CODE1 SEGMENT

START: MOV   AX,SEG DATA1

             MOV   DS,AX

             LEA   DX,MSG

             MOV   AH,09H

             INT 21H

             LEA   DX,NUMBER

             MOV   AH,0AH

             INT   21H

             LEA   BX,NUMBER+4

             MOV   AH,[BX]

             MOV   AL,[BX+1]

             AAD

             MOV   BL,04H

             DIV   BL

             AND   AH,0FFH

             JZ   YES

             LEA   DX,N

             MOV   AH,09H

             INT   21H

             JMP   DOWN

YES: LEA   DX,YS

             MOV   AH,09H

             INT   21H

```
DOWN:       MOV       AH,4CH
            INT       21H
CODE1       ENDS
END         START
```

16. 8086 Program On Fibonacci Series.

```
DATA SEGMENT

DATA ENDS

CODE SEGMENT

ASSUME  DS:DATA, CS:CODE

MAIN PROC    MOV      AH,02H

             MOV    DL,'1'

             INT    21H

             MOV    DL, '1'

             INT    21H

             MOV    DL,'1'

             INT    21H

             MOV    BL,01H

             MOV    CH,01H
```

```
            MOV   DL,'1'

            INT  21H

START1:    MOV    CL,BL

            ADD   BL,CH

            MOV   CH,CL

            MOV   AL,BL

            MOV   AH,00

            MOV   BH,10

            DIV   BH

            MOV   CL,AH

            MOV   DL,AL

            MOV   AH,02H

            ADD   DL,30H

            INT   21H

            MOV   DL,CL

            ADD   DL,30H

            INT   21H

            MOV   DL,CL
```

```
            ADD    DL,30H

            INT    21H

            MOV    DL,'1'

            INT    21H

            CMP    BL,50H

JL START1 :  MOV    AX,4C00H

            INT    21H

CODE        ENDS

END
```

17. 8086 Program to transfer a given block of data from source memory block to destination memory block without overlap.

DATA SEGMENT

VAR1   DW 12H,34H,45H,67H,56H

CNT   DW 5

RES   DW ?

DATA ENDS

CODE SEGMENT

```
ASSUME CS:CODE,DS:DATA

START:    MOV   AX,DATA

          MOV   DS,AX

          MOV   AX,CNT

          MOV   SI,0000H

NEXT:     MOV   AX,VAR1[SI]

          MOV   RES[SI],AX

          INC   SI

          INC   SI

          LOOP  NEXT

          MOV   AH,4CH

          INT   21H

CODE   ENDS

END    START*****************
```

18. 8086 Program To Perform Reverse a String.

DATA SEGMENT

M1 DB 10,13,'ENTER THE STRING:$'

M2 DB 10,13,'REVERSE OF A STRING:$'

BUFF DB 80DB 0DB 80 DUP(0)

COUNTER1 DW 0

COUNTER2 DW 0

DATA ENDS

CODE SEGMENT

ASSUME CS: CODE, DS:DATA

```
START:      MOV     AX,DATA

            MOV     DS,AX

            MOV     AH,09H

            MOV     DX,OFFSET M1

            INT     21H

            MOV     AH,0AH

            LEA     DX,BUFF

            INT     21H

            MOV     AH,09H

            MOV     DX,OFFSET M2

            INT     21H

            LEA     BX,BUFFINC BX
```

```
            MOV    CH,00

            MOV    CL,BUFF+1

            MOV    DI,CX

BACK:       MOV    DL,[BX+DI]

            MOV    AH,02H

            INT    21H

            DEC    DI

            JNZ    BACK

EXIT:       MOV    AH,4CH

            INT    21H

CODE    ENDS

END     START
```

19.   Accepts a string and a character from the user and displays the count and the position of the character.

MODEL SMALL

DISPLAY MACRO MSG

```
    MOV    DX,OFFSET MSG

    CLD

    MOV    AH,09H

    INT    21H

ENDM

DATA

BUFFER DB 50H

DB ?

DB 50H DUP(?)

POSITION_ARRAY DB 50H DUP(?)

POSITION_ARRAY_DISPLAY DB 50H DUP(?)

COUNT DB ?

DB ?

DB '$'

MSG1 DB 10,13," ENTER A STRING : $"

MSG2 DB 10,13," ENTER THE CHARACTER : $"
```

MSG3 DB 10,13," THE COUNT OF THE GIVEN CHARACTER IS : $"

MSG4 DB 10,13," AND THE POSITION(S) ARE : $"

MSG5 DB 10,13," CHARACTER NOT FOUND $"

CODE

```
    MOV    AX,DATA

    MOV    DS,AX
```

DISPLAY MSG1

```
    MOV    DX,OFFSET BUFFER

    CLD

    MOV    AH,0AH

    INT    21H
```

DISPLAY MSG2

```
    CLD

    MOV    AH,01H

    INT    21H
```

```
        MOV     BL,AL ; CHARACTER IN BL

        MOV     BH,BUFFER+1 ; STRING LENGTH IN BH

        XOR     CX,CX

        LEA     DI,POSITION_ARRAY

        LEA     SI,BUFFER+2

        MOV     CL,01H ; POSITION IN CL REGISTER

UP: CMP     BL,BYTE PTR[SI]

JNZ NEXT

        INC     CH ; INCREMENT COUNT (COUNT IN CH

                REGISTER )

        MOV     BYTE PTR[DI],CL

        INC DI

NEXT: INC CL

        INC     SI

        DEC     BH

        JNZ     UP
```

```
CMP     CH,01H

JB   NOTFOUND

MOV     AL,CH

CALL    HEX_TO_ASCII

MOV     COUNT,AH

MOV     COUNT+1,AL

DISPLAY MSG3

DISPLAY     COUNT

MOV     DH,CH

LEA     SI,POSITION_ARRAY

LEA     DI,POSITION_ARRAY_DISPLAY

UP2: MOV    AL,BYTE PTR[SI]

CALL    HEX_TO_ASCII

MOV     BYTE PTR[DI],AH

INC     DI

MOV     BYTE PTR[DI],AL
```

```
INC     DI

MOV     BYTE PTR[DI],20H

INC     DI

INC     SI

DEC     DH

JNZ     UP2

MOV     BYTE PTR[DI],'$'

DISPLAY MSG4

DISPLAY POSITION_ARRAY_DISPLAY

JMP EXIT

HEX_TO_ASCII PROC NEAR

PUSH CX

XOR     CX,CX

MOV     AH,AL

MOV     CL,04H

ROR     AH,CL
```

```
        AND     AH,0FH

        CMP     AH,09H

        JBE     H1

        ADD     AH,37H

        JMP     H2

H1: ADD         AH,30H

H2: AND         AL,0FH

        CMP     AL,09H

        JBE     NEXT2

        ADD     AL,37H

        JMP     NEXT3

NEXT2: ADD      AL,30H

NEXT3: POP      CX

        RET

        NOTFOUND: DISPLAY MSG5

EXIT: MOV AH,04CH
```

INT 21H

END

20. 8086 Program of 10's compliment of an 8 bit number.

MOV   AH,0B

MOV   AL,[1100]

SUB   AH,AL

MOV   [1200],AH

HLT

21. 8086 Programs for numbers of +ve & -ve numbers.

JNZ   L2

JMP   L3

L1 :   INC   SI

INC   AH

DEC   CX

JNZ   L2

L3:   HLT

22. 8086 Program for square root of a number.

```
        MOV     AX,[1100]

        MOV     BX,0FFFF

        MOV     CX,00

L1: INC         CX

        ADD     BX,02

        SUB         AX,BX

        JNZ     L1

        MOV     [1200],CX

HLT
```

23. 8086 Programs to separate Even and Odd parity.

```
        MOV     SI,1200

        MOV     DI,1300

        MOV     BX,1400

        MOV     CX,0A

        MOV     AX,[SI]

L3: MOV     AX,[SI]
```

```
     ADD     AX,00

     JPE     L1

     MOV     [DI],AX

     JMP L2

L1: MOV      [BX],DI

     INC     BX

L2: INC         SI

     DEC         CX

     JNZ     L3

HLT
```

CHAPTER VI

MINIMUM MODE OF 8086

MINIMUM MODE 8086 SYSTEM AND TIMINGS

In a minimum mode 8086 system, the microprocessor 8086 is operated in minimum mode by strapping its MN/MX* pin to logic1. In this mode, all the control signals are given out by the microprocessor chip itself. There is a single microprocessor in the minimum mode system. The remaining components in the system are latches, transreceivers, clock generator, memory and I/O devices. Some type of chip selection logic may be required for selecting memory or I/O devices, depending upon the address map of the system.

The latches are generally buffered output D-type flip-flops, like, 74LS373 or 8282. They are used for separating the valid address from the multiplexed address/data signals and are controlled by the ALE signal generated by 8086. Transreceivers are the bidirectional buffers and some times they are called as data amplifiers. They are required to separate the valid data from the time multiplexed address/data signal. They are controlled by two signals, namely, DEN* and DT/R*. The DEN* signal indicates that the valid data is available on the data bus, while DT/R indicates the direction of data, i.e. from or to the processor. The system contains memory for the monitor and users program storage. Usually, EPROMS are used for monitor storage, while RAMs for users program storage. A system may contain I/O devices for communication with the processor as well as some special purpose I/O devices. The clock generator generates the clock from the crystal oscillator and then shapes it and divides to make it more precise so that it can be used as an accurate timing reference for the system. The clock generator also synchronizes some external signals with the system clock. The general system organization is shown in Fig 6.1. Since it has 20 address lines and 16 data lines, the 8086 CPU requires three octal address latches and two octal data buffers for the complete address and data separation. The working of the minimum mode configuration system can be better described in terms of the timing diagrams rather than qualitatively describing the operations. The op code fetch and read cycles are similar. Hence the timing diagram can be categorized in two parts, the first is the timing diagram for read cycle and the second is the timing diagram for write cycle.

Fig 6.2 shows the read cycle timing diagram. The read cycle begins in T1 with the assertion of the address latch enable (ALE) signal and also M/IO* signal. During the negative going edge of this signal, the valid address is latched on the local bus. The BHE* and A0 signals address low, high or both bytes. From Tl to T4, the M/IO* signal indicates a memory or I/O operation. At T2 the address is removed from the local bus and is sent to the output. The bus is then tristated. The read (RD*) control signal is also activated in T2 .The read (RD) signal causes the addressed device to enable its data bus drivers. After RD* goes low, the valid data is available on the data bus. The addressed 26 device will drive the READY line high, when the processor returns the read signal to high level, the addressed device will again tri state its bus drivers.

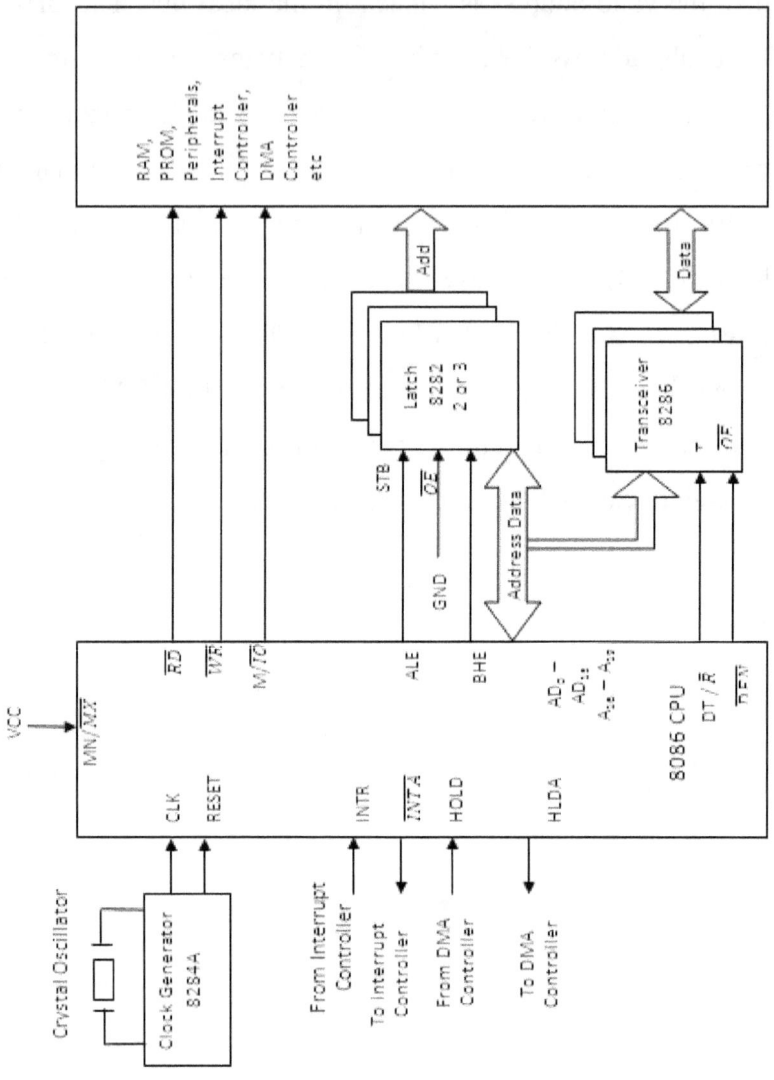

Fig 6.1 General system organization

Fig 6.3 shows the write cycle timing diagram. A write cycle also begins with the assertion of ALE and the emission of the address. The M/$\overline{IO}$ signal is again asserted to indicate a memory or I/O operation. In T2 after sending the address in T1 the processor sends the data to be written to the addressed location. The data remains on the bus until middle of T4 state. The $\overline{WR}$ becomes active at the beginning ofT2 (unlike $\overline{RD}$ is somewhat delayed in T2 to provide time for floating).

The $\overline{BHE}$ and A0 signals are used to select the proper byte or bytes of memory or I/O word to be read or written. The M/$\overline{IO}$, $\overline{RD}$ and $\overline{WR}$ signals indicate the types of data transfer as specified in Table

M/IO	RD	WR	Transfer	Type
0	0	1	I/O	read
0	1	0	I/O	write
1	0	1	Memory	read
1	1	0	Memory	write

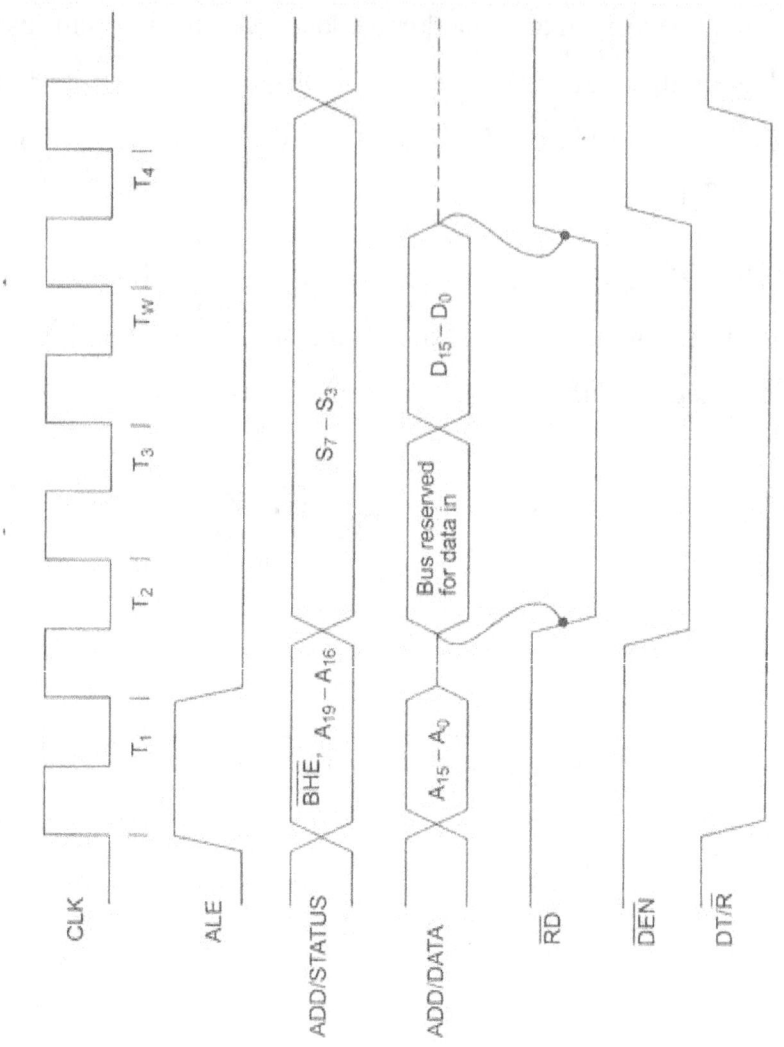

Fig.6.2. Read Cycle Timing Diagram for Minimum Mode

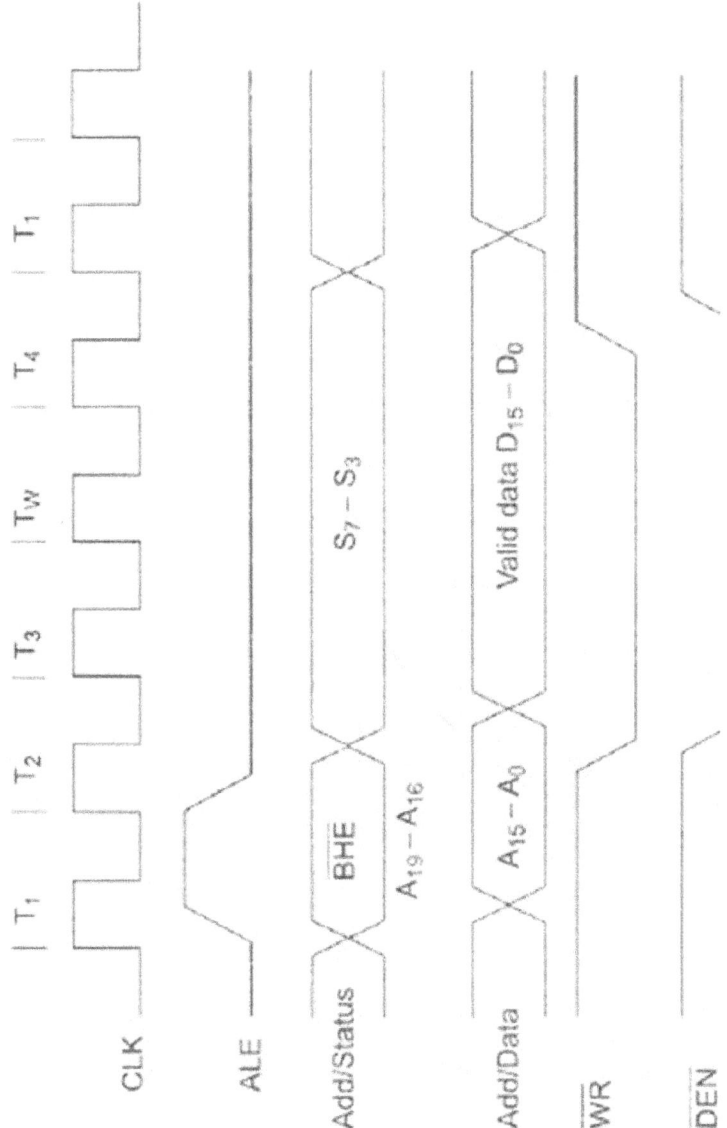

Fig 6.3 Write Cycle Timing Diagram for Minimum Mode

HOLD Response Sequence

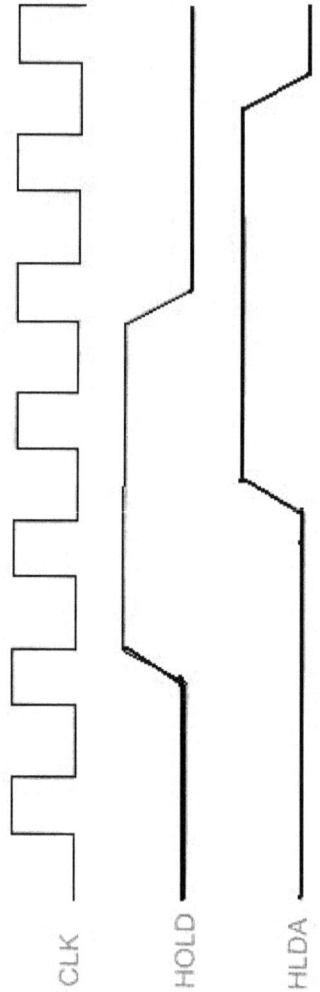

Fig 6.4. Bus Request and Bus Grant Timings in Minimum Mode System

The HOLD pin is checked at the end of the each bus cycle. If it is received active by the processor before T4 of the previous cycle or during T1 state of the current cycle, the CPU activities HLDA in the next clock cycle and for the succeeding bus cycles, the bus will be given to another requesting master The control control of the bus is not regained by the processor until the requesting master does not drop the HOLD pin low. When the request is dropped by the requesting master, the HLDA is dropped by the processor at the trailing edge of the next clock as shown in fig 6.4.

MAXIMUM MODE 8086 SYSTEM AND TIMINGS

In the maximum mode, the 8086 is operated by strapping the MN/$\overline{MX}$* pin to ground. In this mode, the processor derives the status signals $\bar{s}_2$, $\bar{s}_1$ and $\bar{s}_0$. Another chip called bus controller derives the control signals using this status information. In the maximum mode, there may be more than one microprocessor in the system configuration. The other components in the system are the same as in the minimum mode system. The general system organization is as shown in the fig 6.1

The basic functions of the bus controller chip IC8288, is to derive control signals like $\overline{RD}$ and $\overline{WR}$ (for memory and I/O devices), $\overline{DEN}$, DT/$\overline{R}$, ALE, etc. using the information made available by the processor on the status lines. The bus controller chip has input lines $\bar{s}_2$, $\bar{s}_1$ and $\bar{s}_0$ and CLK. These inputs to 8288 are driven by the CPU. It derives the outputs ALE, $\overline{DEN}$, DT/$\bar{R}$, $\overline{MWTC}$, $\overline{AMWC}$, $\overline{IORC}$, $\overline{IOWC}$ and $\overline{AIOWC}$. The $\overline{AEN}$, IOB and CEN pins are specially useful for multiprocessor systems. $\overline{AEN}$ and IOB are generally grounded. CEN pin is usually tied to +5V.

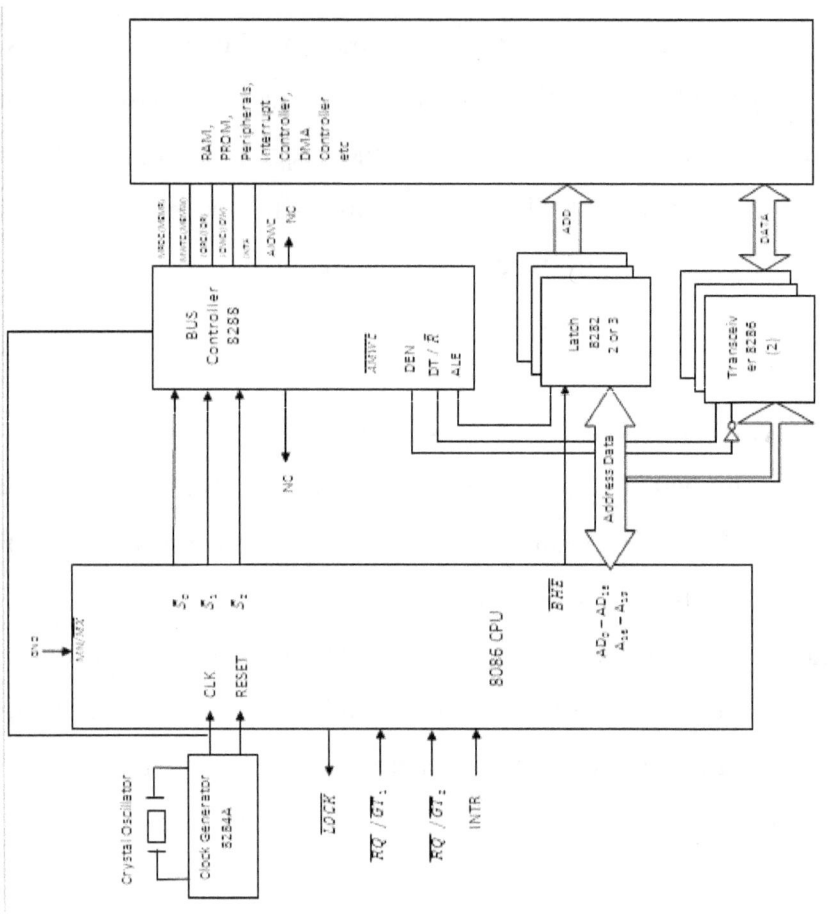

Fig 6.5 Maximum Mode 8086 System

The significance of the MCE/$\overline{PDEN}$ output depends upon the status of the IOB pin. If IOB is grounded, it acts as master cascade enable to control cascaded 8259A; else it acts as peripheral data enable used in the multiple bus configurations. $\overline{INTA}$ pin is used to issue two interrupt acknowledge pulses to the interrupt controller or to an interrupting device.

$\overline{IORC}, \overline{IOWC}$ are I/O read command and I/O write command signals respectively. These signals enable an IO interface to read or write the data from or to the addressed port. The $\overline{MRDC}, \overline{MWTC}$ are memory read command and memory write command signals respectively and may be used as memory read and write signals. All these command signals instruct the memory to accept or send data from or to the bus. For both of these write command signals, the advanced signals namely $\overline{AIOWC}$ and $\overline{AMWTC}$ are available. They also serve the same purpose, but are activated one clock cycle earlier than the $\overline{IOWC}$ and $\overline{MWTC}$ signals, respectively. The maximum mode system is shown in fig. 6.5.

The maximum mode system timing diagrams are also divided in two portions as read (input) and write (output) timing diagrams. The address/data and address/status timings are similar to the minimum mode. ALE is asserted in T1, just like minimum mode. The only difference lies in the status signals used and the available control and advanced command signals. The fig. 6.7 shows the maximum mode timings for the read operation while the fig. 6.8 shows the same for the write operation.

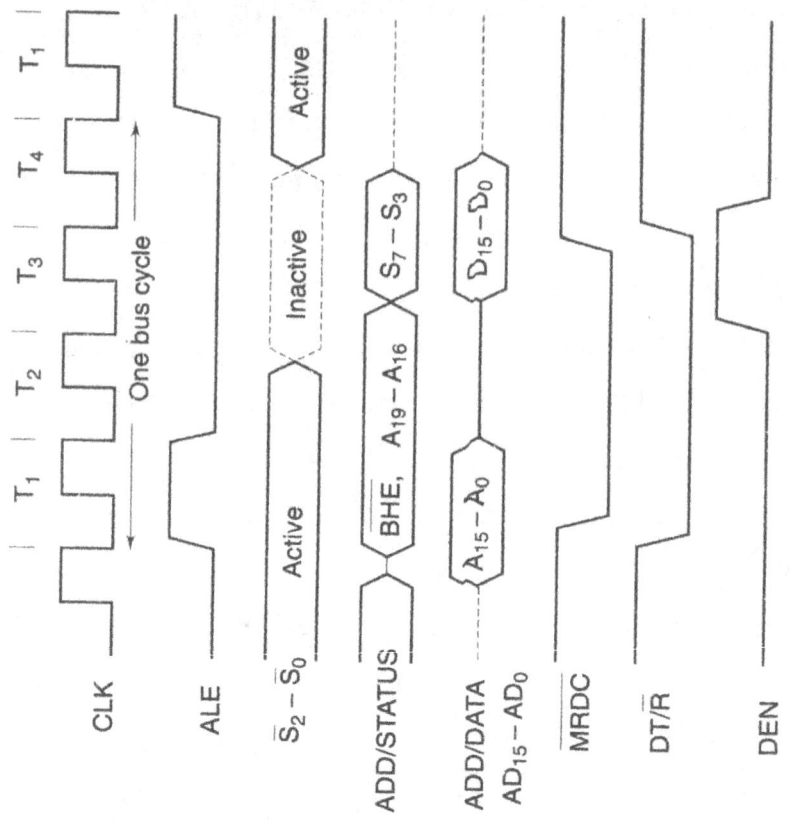

Fig. 6.7 Memory Read Timing in Maximum Mode

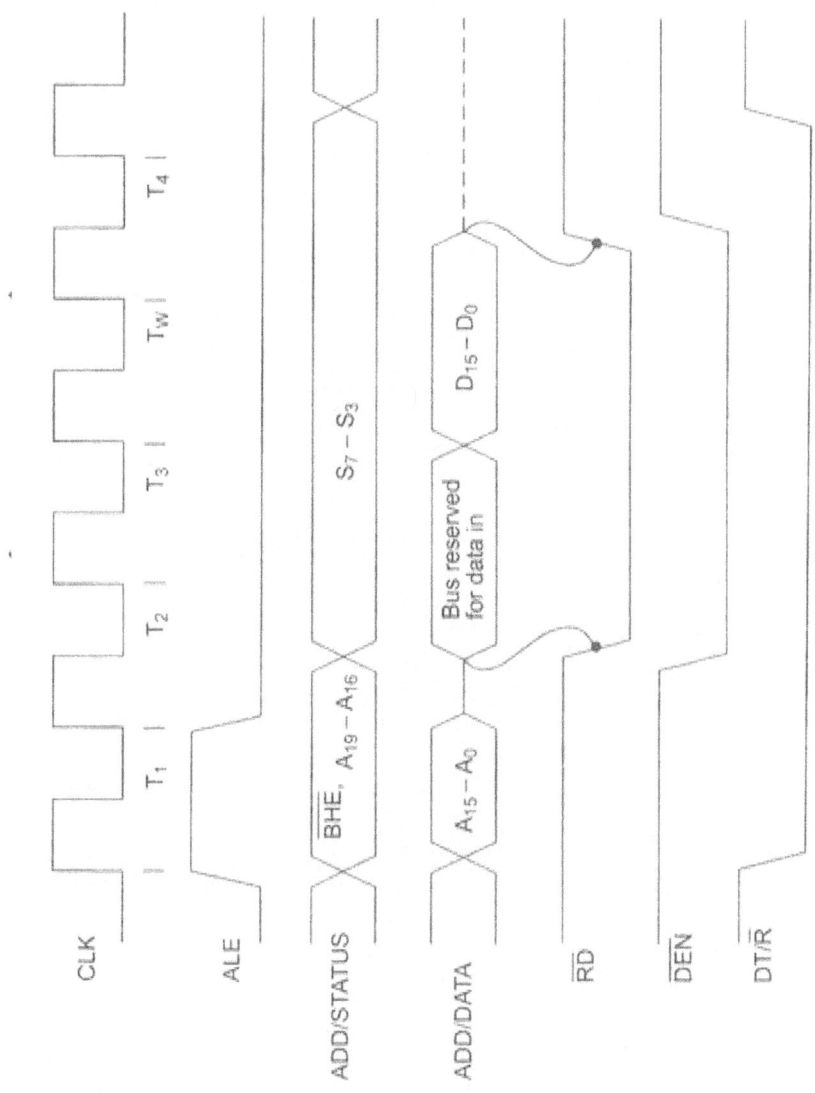

Fig. 6.8 Memory Write Timing in Maximum Mode

Timings for RQ*/GT* Signals

The request/grant response sequence contains a series of three pulses. The request/grant pins are checked at each rising pulse of clock input. When a request detected the processor issues grant pulse over RQ*/GT* pin immediately during T4 (current) or T1(next) state. When the requesting master receives this pulse, it accepts the control the control of the bus. The requesting master uses the bus till it requires. When it is ready to hand over. The bus, it sends a release pulse to the processor (host) using RQ/GT pin. The sequence is shown in fig 6.9.

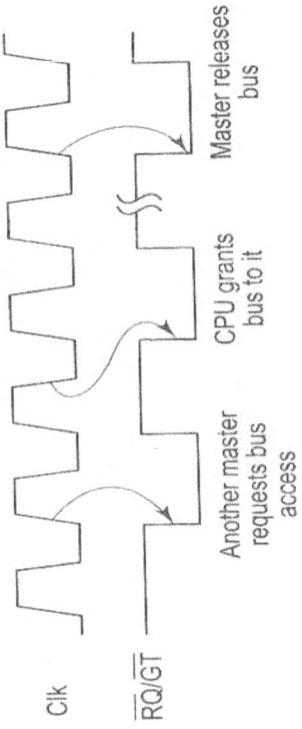

Fig 6.9 RQ*/GT* Timings in Maximum Mode

CHAPTER VII

Architecure of 80286

SALIENT FAETURES OF 80286

1. The 80286 has 24-bit address bus and hence it can be able to handle 16 Mbyte of physical memory.

2. The clock frequency for different versions are 12.5 MHz ,10 MHz and 8 MHz.

3. Intel's 80286 is the first CPU to incorporate the integrated memory management unit.

4. 80286 introduces the concept of virtual memory.

5. Memory management provides data protection or unauthorised access prevention.

6. The 80286 has 16 bit data bus.

7. 80286 works in two operating modes viz: real address mode and protected virtual address mode.

8. 80286 is upwardly compatible with 8086 in terms of instruction set.

9. 80286 CPU contains almost the same set of register as in 8086.

REGISTER ORGANISATION OF 80286

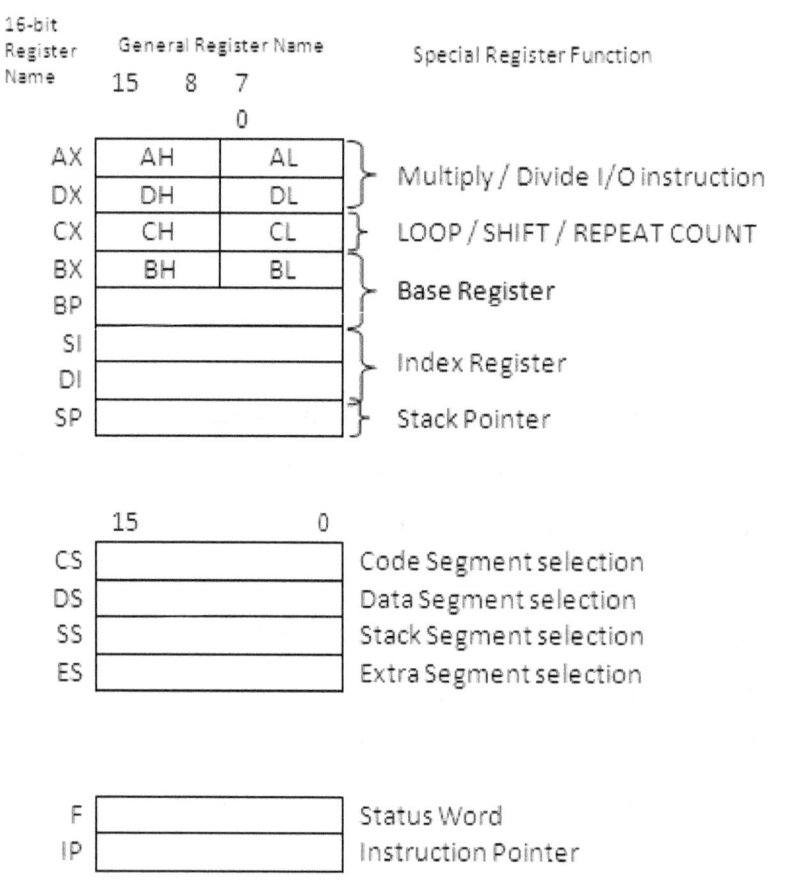

D_{15}	D_{14}	D_{13}	D_{12}	D_{11}	D_{10}	D_9	D_8	D_7	D_6	D_5	D_4	D_3	D_2	D_1	D_0
X	NT	IOPL		OF	DF	IF	TF	SF	ZF	X	AF	X	PF	X	CF

D_{31}	D_{30}	D_{29}	D_{28}	D_{27}	D_{26}	D_{25}	D_{24}	D_{23}	D_{22}	D_{21}	D_{20}	D_{19}	D_{18}	D_{17}	D_{16}
X	X	X	X	X	X	X	X	X	X	X	X	TS	EM	MP	PE

X : Not used /reserved bits

CF : Carry flag

PF : Parity flag

AF : Auxiliary carry flag

ZF : Zero flag

SF : Sign flag

TF : Trap flag

IF : Interrupt flag

DF : Direction flag

NT : Nested Task

IOPL: I/O Privilege level

PE : protection Enable

MP : Monitor Processor Extension

EM :   Processor Extension Emulator

TS :   Task Switch

The 80286 CPU contains almost the same set of registers as in 8086.

1. Eight 16 bit general purpose register
2. Four 16 bit segment register
3. Status and control register
4. Instruction pointer

General purpose register

There are four 16-bit general purpose registers AX, BX, CX & DX. Each of these 16 bit registers are further subdivided into two 8-bit registers as AL, AH, BL, BH, CL, CH, DL, DH. The letters L & H specify the lower and higher byte of a particular register.

Register AX servers as accumulator register BX servers as base register for computation of memory address. Register CX used as a counter register in case of multi iteration instruction. Register DX is used for memory addressing when data are transferred between I/O port & memory using certain I/O Instruction.

Pointers and Index register

The following four registers are the group of pointers and index registers

e) Stack pointer

f) Base pointer

g) Source index

h) Destination index

The pointers contains offset within the segment. Base pointer contains offset within the data segment. The function of Source index contain offset within stack segment. The function of source pointer is same as stack pointer in 8085.

The index registers are used as general purpose register as well as for offset storage in case of Indexed, base indexed & relative base indexed addressing modes. Register SI is used to store the offset of source data in data segment while register destination index is used to store the offset of destination in data or extra segment.

Segment register

There are four segment register

e. Code segment register

f. Data segment register

g. Stack segment register

h. Extra segment register

The 8086 address the segmented memory. The complete 1Megabytes memory is divided into 16-bit logical segments. Each segment thus contains 64 Kilobytes of memory. The segment register of 8086 acts as base register. The segment register point out the starting memory address of the currently used segments.

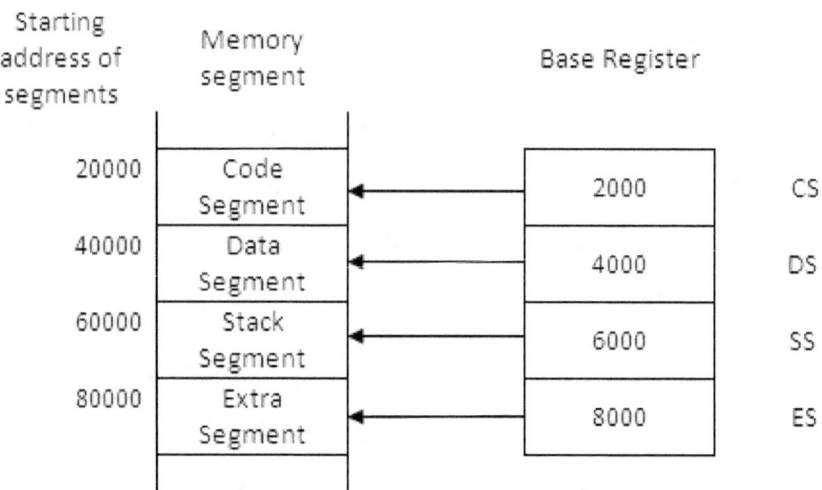

The code segment register is used for address the memory location in the code segment (CS) of the memory where the executable program is stored. The data segment register points to the data segment of the memory where the data resides. The extra segment register also contains data. The stack segment register is used for addressing the memory where the stack data is stored.

Instruction pointer (IP)

The instruction pointers contents the offset within the code segment. The instruction pointer acts as a program counter. It indicates the next instruction to be executed. Its contents are automatically incremented when the execution of the program proceeds further.

INTERNAL BLOCK DIAGRAM OF 80286

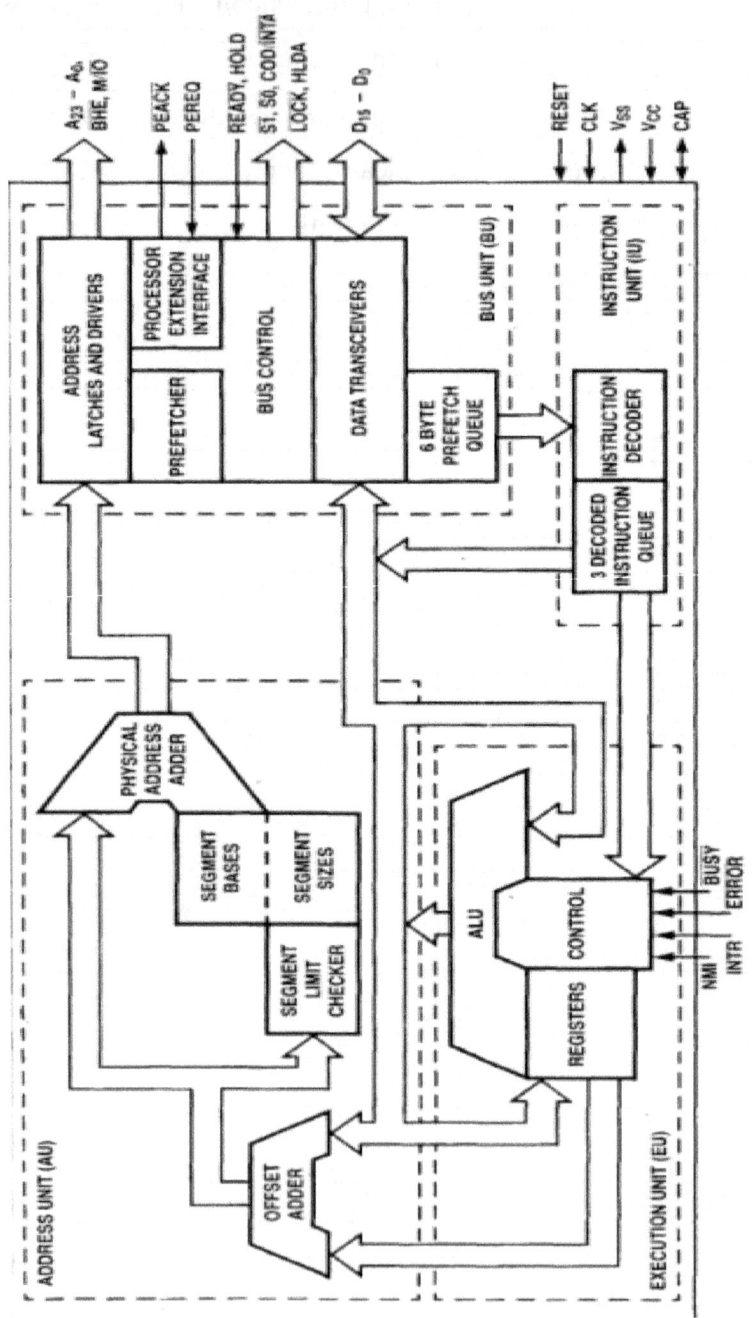

The internal block diagram of 80286 CPU has 4 functional parts:

1. Address Unit (AU)

2. Bus Unit (BU)

3. Instruction Unit (IU)

4. Execution Unit (EU)

The address unit is responsible for calculating the address of instruction and data that the CPU wants to access. The address line derived by this unit may be used to address different peripherals.

This physical address computed by AU is handed over to Bus Unit (BU) of CPU. The address latches & drives in BU transmits the physical address to address bus $A_{23} - A_0$. The instructions are fetched from memory in advance and stored in a queue to enable faster execution of instructions. This concept is known as instruction pipelining. The pre-fetcher module in BU perform this task of pre-fetching. The bus control module in BU control pre-fetcher module. These fetched instructions are arranged in 6 byte pre-fetch queue. For branch instructions pre-fetcher will flush out the queue and further (branched) pre-fetching may carry out. Another major module in the bus unit is the processor extension interface module which takes care of communication between CPU and CO-processor.

The 6 byte pre-fetch queue forwards the instructions to Instruction Unit (IU) . The instruction unit accepts instructions from the pre-fetch queue and an instruction decoder decodes them one by one. The data transreceiver interface and control the internal data bus with system bus.

The output of decoding circuit drives a control circuit in the Execution Unit(EU) which is responsible for executing the instructions received from the decoded instruction queue which sends the data part of the instruction over the data bus. The EU contains the register bank, used for storing the data as scratch pad or used as special purpose register. The ALU – the heart of the EU- carries out all the arithmetic and logical operations and sends the results either over the data bus or back to the register bank.

ADDRESSING MODES OF 80286

The 80286 supports eight addressing modes to access the operands stored in memory.

1. Register operand mode:

The operand in this mode is located in one of the 8 bit or 16 bit general purpose register.

2. Immediate operand mode:

In this mode , the immediate operand includes in the instruction itself. In the remaining six addressing modes, the operand is located in a memory segment. A memory operand address in these modes may be computed using two 16 bit components : segment selector and offset. The different combinations of immediate displacement, base register, pointers and index register result in the following six operating modes:

3. Direct mode :

The offset is a part of instruction either as 8 bit or 16 bit immediate operand(displacement)

4. Register mode :

The operand is stored either in any of the general purpose register or in SI, DI ,BX ,or BP.

5. Based mode :

The offset is obtained by adding a displacement and the contents of one of the base registers, either BX or BP.

6 .Indexed mode :

The offset is obtained by adding a displacement with the contents of an index register, either SI or DI.

7 . Based Indexed mode :

The operand is stored at a location whose address is calculated by adding the contents of any of the base register with the contents of any of the index register.

8. Based Indexed mode with displacement

In this mode, the offset of the operand is calculated by adding an 8 bit or 16 bit immediate displacement with the contents of a base register and an index register.

INTERRUPTS OF 80286

The interrupts of 80286 may be divided into 3 categories:

1.External/Internal interrupts

2.INT instruction/software interrupts

3.Interrupts generated internally by exception

While trying to execute a divide by zero instruction , the CPU detects a major error and stops further execution. In this case, we say that an exception has been generated.

As in 8086, the 80286 interrupt vector table requires 1 Kbytes of space for storing 256, 4-byte pointers to point to the corresponding 256 ISR (Interrupt Service Routine).

Each pointer contains 16 –bit offset followed by 16 bit segment selector to point to a particular ISR. The INT type interrupts (INT 00H to INT FFH) are similar to 8086.

In external interrupt, an external device or a signal interrupts the processor from outside viz keyboard interrupt, NMI pin. The internal interrupt is generated internally by the processor circuit or by the execution of interrupt instruction viz divide by zero interrupt, overflow interrupt, interrupt due to INT instructions TRAP etc.

In 8086 microprocessor there are two interrupt pins-NMI(non-maskable interrupt) and INTR(interrupt request).

NMI has the highest priority among the external interrupt.ie NMI input cannot be masked or disable by any means.

INTR has lower priority as compared to NMI. The INTR interrupt may be masked using interrupt flag(IF).The INTR is of 256 types. The INTR types may be from 00H to FFH(0 to 255).If more than one type of INTR interrupt occurs at a time, then an external chip called programmable interrupt controller is required to handle them. IF is SET, the processor is ready to respond to any INTR interrupt .if the IF is reset, the processor will not serve any interrupt appearing at this pin. However, once the processor responds to an INTR signal, the IF is automatically RESET .

TRAP(single set -type1) is an internal interrupt having the highest priority amongst all interrupt except divide by zero(Type 0) exception.

PIN DISCRIPTION OF 80286

PEREQ and $\overline{PEACK}$:

(Processor Extension Request and Acknowledgement)

Processor extension refers to co-processor (80287 in case of 80286 CPU). This pair of pins extend the memory management and protection capabilities of 80286 to the processor extension 80287. The PEREQ input requests the 80286 to perform a data operand transfer for a processor extension .The $\overline{PEACK}$ active low output indicates to the processor extension that the requested operand is being transferred.

$\overline{READY}$:

This active low input pin is used to insert wait states in a bus cycle, for interfacing low speed peripherals. This signal is neglected during HLDA cycle.

$\overline{LOCK}$

This active low output pin is used to prevent the other masters from gaining the control of bus for the current and the following bus cycles. This pin is activated by a LOCK instruction prefix or automatically by hardware during XCHG, interrupt acknowledge or descriptor table access.

$\overline{BUSY}$ and $\overline{ERROR}$:

Processor extension $\overline{BUSY}$ and $\overline{ERROR}$ active low input signals indicates the operating conditions of a processor extension to 80286. The $\overline{BUSY}$ goes low, indicating 80286 to suspend the execution & wait until the $\overline{BUSY}$ becomes inactive. In this duration, the processor extension is busy with its allotted job. Once the job is completed the processor extension drives the $\overline{BUSY}$ input high indicating 80286 to continue with program execution.

An active $\overline{ERROR}$ signal causes the 80286 to perform the processor extension interrupt while executing the WAIT & ESC instruction. The active $\overline{ERROR}$ signal indicates to 80286 that the processor extension has committed a mistake & hence it is reactivating the processor extension.

REAL ADDRESSING MODE

The 80286 CPU operates in two modes:
1. Real address mode
2. Protected Virtual Address Mode (PVAM)

In real addressing mode of operation of 80286 acts as a fast 8086.the instruction set is upwardly compatible with that of 8086. The 80286 addresses only 1Mbyte of physical memory using $A_0 - A_{19}$. The lines $A_{20} - A_{23}$ are not used by the internal circuit of 80286 in this mode.

In real address mode , while addressing the physical memory, the 80286 uses BHE along with $A_0 - A_{19}$. The 20 – bit physical address is again formed in the same way as that in 8086. The contents of segment registers are used as segment base addresses. The other registers depending upon the addressing mode contain the offset addresses.

The address formation in real mode is as shown in figure

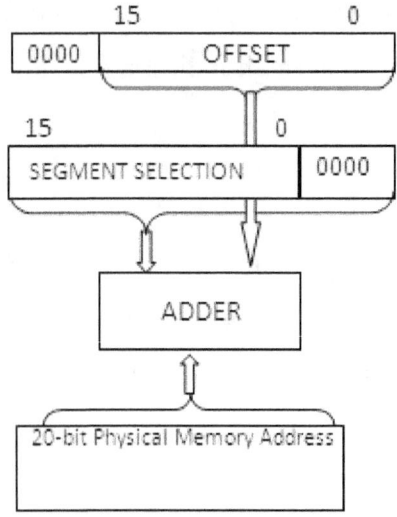

PROTECTED VIRTUAL ADDRESS MODE (PVAM)

The 80286 is the first processor to support the concept of virtual memory and memory management. Though the virtual memory does not exist physically it still appears to be available within the system. The concept of virtual memory is implemented using physical memory that the CPU can directly access and the secondary memory that is used as a storage for data and program, which are stored in secondary memory initially. The segment of the program or data required for actual execution at that instant is fetched from the secondary memory into physical memory. After execution of this fetched segment the next segment required for further execution is again fetched from the secondary memory,while the results of the executed segments are stored back into the secondary memory for further references. This continues till the complete program is executed.

The procedure of fetching the chosen program segments or data from the secondary storage onto physical memory is called swapping . The procedure of storing back the partial results or data back onto the secondary storage is called unswapping. The virtual memory is allotted per task. The 80286 is able to address 1 Gbyte (2 30 bytes) of virtual memory per task. The complete virtual memory is mapped on to 16 Mbyte physical memory.

In case of huge programs (in general greater than physical memory in size) they are divided in either smaller segments or pages which are arranged in appropriate sequence and are swapped. These segments or pages have been associated with a data structure called as descriptor. The set of the descriptor may be called descriptor table.

The descriptor contains information of the program segment or page. Corresponding to different types of segments there may be different types of descriptors.

Viz: For data segment -- data segment descriptor

For code segment – code segment descriptor

For subroutine and interrupt routines – gate descriptor

PHYSICAL ADDRESS CALCULATION IN PVAM

In PVAM the 80286 uses 16 bit contents of a segment register as a selector to address a descriptor stored in physical memory. The segment base address is 24 bit pointer that addresses the first location in that segment . This 24 bit segment base address is added with 16 bit offset to calculate 24 bit physical address. The maximum segment size will be of 64 Kbytes since the offset is only of 16 bits.

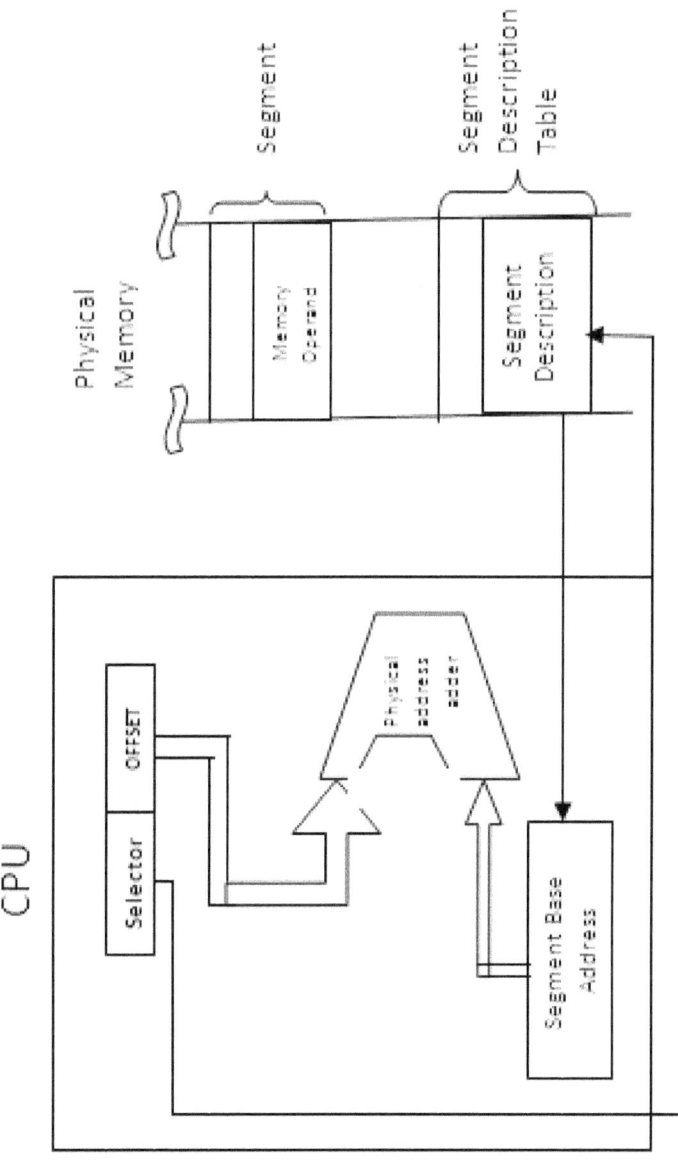

The descriptor is a block of contiguous memory locations containing information of a segment like segment base address, segment limit, segment type , privilege level , segment availability in physical memory , descriptor type and segment use by another task. The base address ie. the starting location of a segment is an important descriptor information . The segment limit indicates the maximum size of a segment . each segment has a type and its privilege level, which indicates the importance of the segment. A certain segment may or may not be present in physical memory at a given time instant, this type of information is also stored in descriptor.

CHAPTER VIII

Architecure of 80386

SALIENT FEATURES OF 80386 DX

1. The 80386 DX is 32 –bit processor that supports 8-bit/16bit/32-bit data operands.

2. 80386 has address bus of 32-bits and hence can access physical memory of 4 Gbytes.

3. The 80386 CPU supports 16K number of segments and thus the total virtual memory space is 4Gbytes *16 K =64 Terrabytes

4. The memory management section of 80386 supports the virtual memory, paging and four levels of protection ,maintaining full compatibility with 80286.

5. The concept of paging is introduced in 80386.

6. 80386 can be supported by 80387 for mathematical data processing.

7. It offers a set of total eight debug registers $DR_0 - DR_7$ for hardware debugging and control.

8. 80386 has on chip address translation cache.

9. 80386 DX is available in 132 pin grid array package and has 20 MHz and 33MHz versions.

DIFFERENCE BETWEEN 80386 DX AND 80386 SX

The 80386 DX is available in another version 80386 SX which has identical architecture as 80386

DX with the difference that it has only 16 bit data bus and 24 bit address bus.

INTERNAL BLOCK DIAGRAM OF 80386

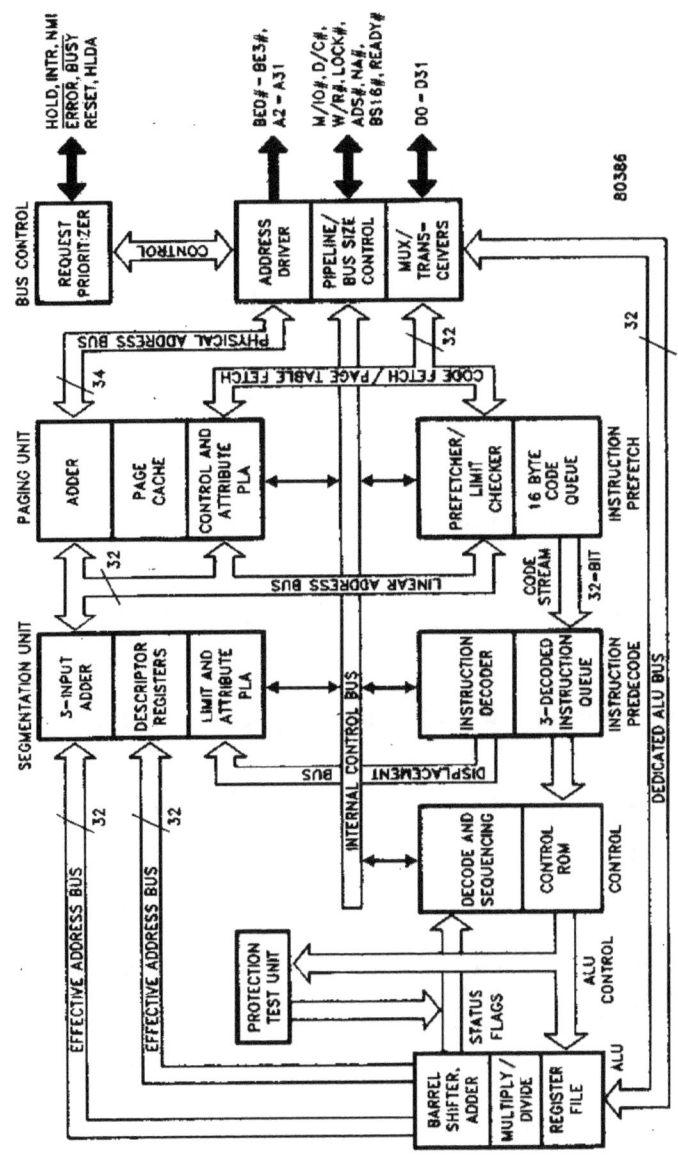

The internal architecture of 80386 is divided into 3 sections :

1. Central Processing Unit (CPU)
2. Memory Management Unit (MMU)
3. Bus Interface Unit (BIU)

The CPU is further divided into execution unit and instruction unit. The execution unit has eight general purpose and eight special purpose registers which are used for handling data or calculating offset addresses . the instruction unit decodes the op-code bytes received from 16 byte instruction code queue and arranges them into 3 instruction decoded instruction queue. Then these instructions are pass to the control section.

The barrel shifter increases the speed of all shift and rotate operations . the multiply/divide logic implements the bit – shift – rotate algorithms to complete the operations in minimum time.

The Memory Management Unit (MMU) consists of a segmentation unit and paging unit. The unit allows the use of two address components viz: segment and offset, for relocability and sharing of code and data. The segmentation unit allows a maximum size of 4 Gbytes segments. The paging unit organises the physical memory in terms of pages of 4 Kbytes size each. Each segment is divided into pages . the virtual memory is organised by the memory management unit. The segmentation unit provides a 4 level protection mechanism for protecting and isolating the system code and data from those of the application program. The paging unit converts linear addresses into physical addresses. The control and attribute PLA checks the privileges at page level. The limit and attribute PLA checks segment limit and attributes at segment level to avoid invalid accesses to code and data in memory segments.

The bus control unit has a prioritize to resolve the priority of the various bus requests. The address driver drives the bus enable and address signals $A_0 - A_{31}$. the pipeline and dynamic bus sizing units handle the related control signals . the data buffers interface the internal data bus with system bus.

DATA TYPES OF 80386

The 80386 supports following 17 data types:

1. Bit
2. Bit field- A group of 32 –bits (4Bytes)
3. Bit string – A group of maximum 4 Gbytes in length
4. Signed byte
5. Unsigned byte
6. Integer word – signed 16 –bit data
7. Long integer 32-bit signed data represented in 2's complement form
8. Unsigned integer word
9. Unsigned long integer
10. Signed Quad word – A signed 64 bit data or 4 word data
11. Unsigned Quad word – unsigned 64-bit data
12. Offset -16/32 bit displacement

13. Pointer – this consists of a pair of 16 bit selector and 16/32 – bit offset

14. Character –An ASCII equivalent to any of the alphanumeric or control characters.

15. Strings- These are the sequences of bytes , words, or double words (maximum 4 G bytes)

16. BCD – Decimal digits from 0-9 represented by unpacked bytes

17. Packed BCD – This represented two packed BCD digits using a byte ie from 00 to 99

REGISTER ORGANISATION OF 80386:-

```
31          15
16          0
        ┌──────┬──────┐
        │  AX  │ EAX
        │  BX  │ EBX
        │  CX  │ ECX
        │  DX  │ EDX
        │  SI  │ ESI
        │  DI  │ EDI
        │  BP  │ EBP
        │  SP  │ ESP
        └──────┘
      Segment selector Register
            ┌──────┐
            │      │ CS      Code
            │      │ SS      Stack
            │      │ DS ┐
            │      │ ES │    DATA
            │      │ FS │
            │      │ GS ┘
            └──────┘
    Instruction Pointer and Flag Register
    31          15
    16          0
        ┌──────┬──────┐
        │  IP  │ EIP
        │ FLAGS│ EFLAGS
        └──────┴──────┘
```

31	18	17	16	15	14	13	12	11	10	9	8	7	6	5	4	3	2	1	0		
Reserved for Intel	VM	RF	0	NT	IL	OP	OF	DF	IF	TF	SF	ZF	0	AF	0	PF	1	CF	EFLAGS	RF – Resume Flag

RF – Resume Flag

VM – Virtual Mode

Debug Register and Test Registers.

31	0	
Linear Break Point Address 0		DR_0
Linear Break Point Address 1		DR_1
Linear Break Point Address 2		DR_2
Linear Break Point Address 3		DR_3
Intel Reserved		DR_4
Intel Reserved		DR_5
Breakpoint Status		DR_6
Breakpoint Control		DR_7

31	0	
Test Control		TR_6
Test Status		TR_7

Test Register for Page Cache

The 80386 has eight 32-bit general purpose registers which may even may be used as 8-bit or 16-bit registers. A 32-bit register, extended register, is represented by E. The function of AX,BX,CX,DX,SI,DI,BP,SP is some as in 8086.

The 80386 has 6 segment registers CS,SS,DS,ES,FS and GS. The CS and SS are code and stack segments respectively while DS,ES,FS and GS are four data segments registers.

The 80386 has 32-bit flag registers. The bits D_0-D_{15} are same as 80286. Two extra new flags are added in 80386.

1. VM(Virtual mode flag)- This is to be set only when 80386 is in protected virtual mode.

2. RF(Resume flag)-This flag is used with the debug register breakpoints.

IOPL flag bits indicate the privilege level of the current Io operations.

The segment descriptor register of 80386 are available for programmer. They are internally used to store the descriptor information like attributes, limit and base addresses of segments. The six segment registers have corresponding six 73-bit descriptor registers.

The 80386 has three 32-bit control register CR_0, CR_2 and CR_3 to hold global machine status independent of the executed task. The control register CR_1 is reserved for use in future Intel processor.

Four special registers are defined to refer to the descriptor table by 80386-

1)Global descriptor Table(GDT)

2)Interrupt descriptor Table(IDT)

3)Local descriptor Table(LDT)

4)Task state segment descriptor (TSS)

Intel 80386 has debug registers DR_0 to DR_7 for hardware debugging.

Addressing modes of 80386:-

The 80386 supports eleven addressing modes to facilitate efficient execution of higher level language programs. The 80386 has all the addressing modes which were available with 80286. In 80386 32-bit immediate or 32-bit registers operands or displacement can be used. The 80386 has family of scaled modes. In case of scaled , any of the index register values can be multiplexed by a valid scaled factor to obtain the displacement. The valid scale factors are 1,2,4 and 8. The different scaled modes are-

1)**Scaled indexed mode**- In this case the contents of index register are multiplexed by a scale factor that may be added further to get the operand offset .

EX:- MOV   EBX, LIST[ESI*2]

2)**BASED Scaled indexed mode**- In this case the contents of index register are multiplexed by a scale factor and then added to base register to obtain the offset.

EX:- MOV   EBX, [EDI*4] [ECX]

EX:- MOV   EBX, [ESI*2] [ECX]

2)**Based Scaled indexed mode with displacement**-The contents of an index register are multiplexed by a scaling factor and the result is added to a base regiter and a displacement to get the offset of an operand.

EX:- MOV   EAX,   LIST[ESI*2] [EBX+0800]

EX:- MOV   EBX,   LIST[EDI*8] [ECX+0100]

The displacement may be any 8-bit , 16-bit or 32-bit immediate numbers. The base and index register may be any general purpose except Esp.

REAL ADDRESS MODE OF 80386:-

In real address mode, 80386 works as a fast 8086 with 32-bit registers and data type. The addressing techniques, memory size, interrupt handling in this mode of 80386 are similar to real address mode of 80286.

In real address mode, 80386 can address at the most 1Mbyte of physical memory using address lines A_0 to A_{19}. Paging unit is disabled and hence the real addresses are the same as the physical address. To form physical address, segment register contents(16-bit) are shifted left by 4 positions and then added to the 16-bit offset address. The segments in 80386 real mode can be read, written or executed ie no protection is available. The segment can be overlapped or non-overlapped. The interrupt vector table of 80386 has been allocated 1kbyte space. The physical address formation is as shown below.

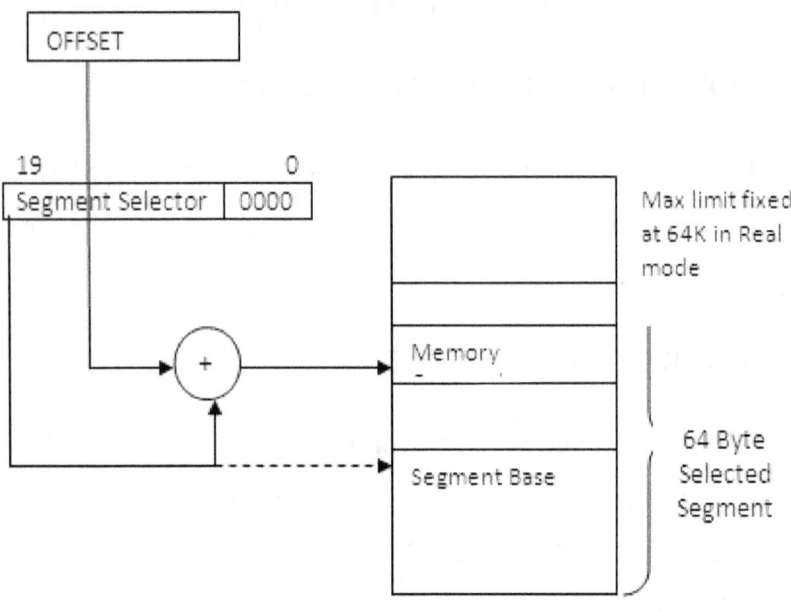

PROTECTED MODE OF 80386:-

All the capabilities of 80386 are available for utilization in its protected mode of operation. In this mode, the 80386 can address 4Gbytes of physical memory and 64 terabytes of virtual memory per task. The 80386 in protected mode supports all software written for 80286 and 8086 to be executed under the control of memory management and protection abilities of 80386.

In this mode, the content of segment registers are used as selectors to address descriptor which contain the segment limit, base address and access rights byte of the segment. The effective address (offset) is added with segment base address to calculate linear address. This linear address is further used as physical address, if the paging unit is disabled otherwise, the paging unit is a memory management unit enabled only in protected mode and allows to handle large segments of memory in terms of 4Kbyte size of pages.

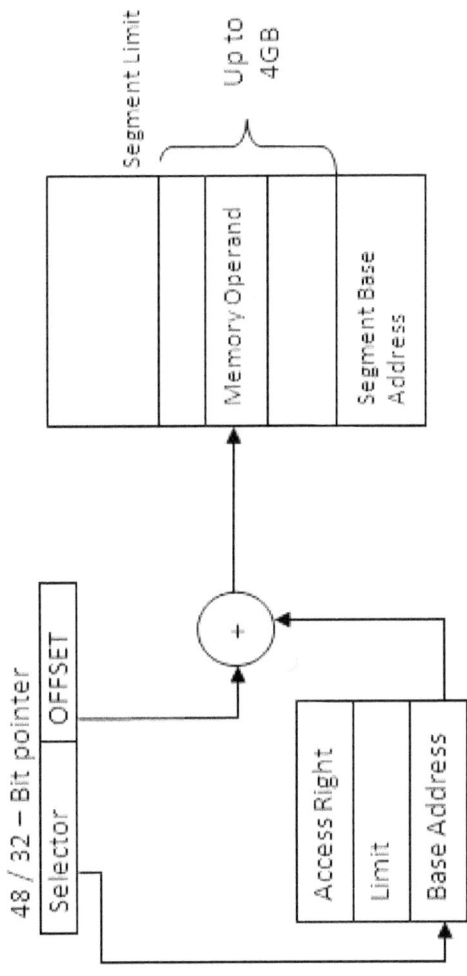

VIRTUAL 8086 MODE:-

Once the 80386 enters the protected mode from the real mode ,it can not return back to the real mode without a reset operation. Thus virtual 8086 mode of operation of 80386, offers an advantage of executing 8086 programs while in protected mode.

The address forming mechanism in virtual mode is exactly mechanism is virtual 8086 mode is exactly identical with that of 8086 real mode. In virtual mode, 8086 can address 1Mbyte of physical memory that may be anywhere in the 4Gbyte address space of the protected mode of 80386. Like 80386,real mode the addresses in virtual 8086 mode lie within 1Mbyte of memory. In the virtual mode ,the paging mechanism and protection capabilities are available. The paging unit allows only 256 pages each of 4Kbytes size. Each of the pages may be located anywhere within the maximum 4Gbyte physical memory. The virtual mode allows the multiprogramming of 8086 applications.

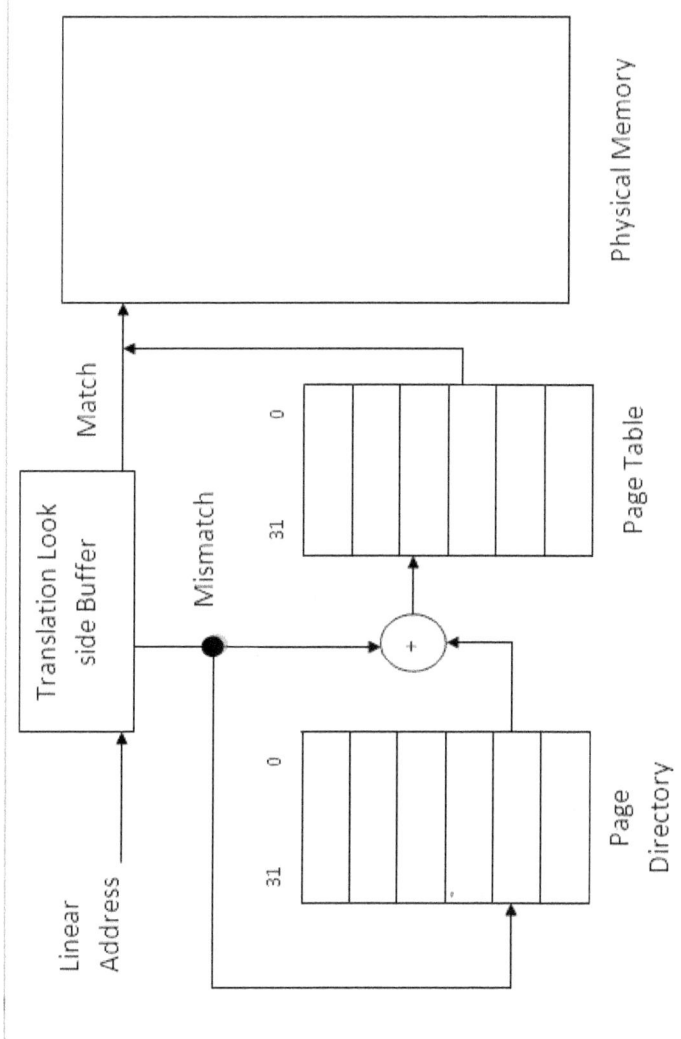

CHAPTER IX

Architecure of 80486

SALIENT FEATURES OF 80846 DX:-

1. 80486 DX is a CPU with a numeric coprocessor.
2. 80486 is 32-bit CPU and is the first CPU within an inbuilt floating-point unit.
3. If retained the complex instruction set (RISC architecture) of 80386 but introduced more pipelining for speed enhancement.
4. It is packaged in 168 pin grid array package.
5. 25 MHz, 35MHz, 50MHz and 100MHz versions are available in 80486.
6. 80486 has introduced five stage pipeline.

7. It is the first processor to have an on chip cache.

INTERNAL BLOCK DIAGRAM / ARCHITECTURE OF 80486:-

The internal architecture of 80486 can be broadly divided into three section

1) Bus interface unit
2) Execution and Control unit
3) Floating point unit

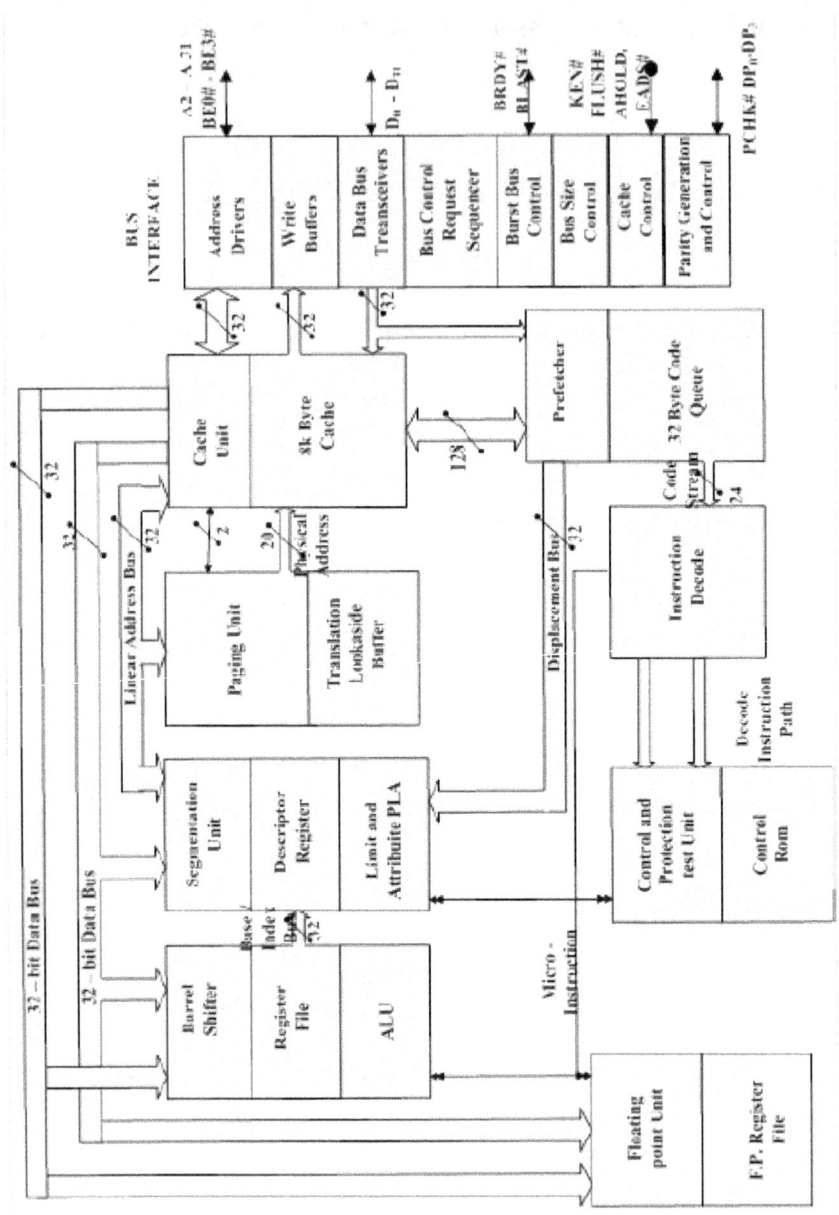

The bus interface unit is mainly responsible for co-ordinating all the bus activities. The address driver interfaces the internal 32-bit address output of cache unit with the system bus. The data bus transreceivers interface the internal 32-bit data bus with the system bus. The 4*80 write data buffer is a queue of four 80-bit registers which hold 80-bit data to be written to the memory. The bus control and request sequence handles bus access and control signals.

The parity generation and control unit maintain the parity and carry out the related checks during the processor operation. The pre-fetches unit fetches the codes from the memory ahead of execution time and arranges them in 32-bit code queue.

The instruction decoder gets code from the code queue and then decodes it sequentially. The output of the decoder drives the control unit to drive the control signals required for the execution of decoded instructions. The protection unit checks if there is any violation of protection forms. The control ROMs stores a microprocessor for driving control signals for execution of different instructions. the register bank and ALU are used for their conventional usages. The barrel shifter helps in implementing the shift and rotate algorithms. The segmentation unit, descriptor register , paging unit, translation look aside buffer and limit and attribute PLA works together to manage the virtual memory of the system and provide adequate protection to the codes or data in the physical memory.

The Floating point units with its register bank communicates with the bus interface unit under the control of memory management unit via its 64-bit internal data bus. The floating-point unit via its 64-bit internal data bus. The floating point unit is responsible for carrying out mathematical data processing at a higher speed as compared to ALU, with its built in floating point algorithms.

REGISTER ORGANISATION OF 80486:-

The register set of 80486 is similar to that of 80386 only a flag called as alignment check(AC) flag is added to the flag register of 80386 at position D_{18}.

If AC Flag bit is set to "1", there is an access to a misaligned and address , a fault is generated.

MODES OF OPERATION OF 80486:-

After reset the 80486, just like 80286 and 80386, start execution in the real address mode. The real address mode of operation of 80486 is exactly the similar to 80386. While executing in real address mode, the 80486 initializes registers, peripherals. IVT sets up descriptor tables and prepares itself for protected mode. The protected mode operation of 80486 is also similar to that of 80386, right from the address formation to descriptor types and structures.

In protected virtual address mode the 80486 also supports a virtual 8086 mode for execution of 8086 application schemas and privilege levels allowed by 80486 are similar to those of 80386. The other operation like task switching, paging and exception handling of 80486 are also similar to 80386.

Salient features of 80586:-

1. 80586 is a CPU with enhanced complex instruction sets which should remain code compatible with earlier X86 CPU is from 8086 to 486.
2. It achieves performance of matching to the third generation RISC performance.
3. It has superscalar, super pipelined architecture.
4. It has two integer pipelines U and V, each one is 4-stage pipeline. This enhances the speed of integer arithmetic of Pentium to a large extent.
5. It has on chips floating point unit which has increased performances of 80386/486 processor.
6. It contains two separate caches viz. data cache and instruction cache.

ENHANCED INSTRUCTION SET OF PENTIUM:-

Besides the instruction of X86 family Pentium also supports computation of several trigonometric and exponential functions through a set of transcendental instructions-

1) FSIN -   to compute sin(Θ)

2) FCOS-   to compute cos(Θ)

3) FSINCOS-   to compute sine and cosine

4) FPTAN-   to compute tan (Θ)

5) FPATAN-   to compute arctan(x)

6) F2XMI -   to compute (2x-1)

7) FYL2X-   to compute y*log2x

8) FYL2XP -   to compute y*log 2(x+1)

Where Θ is an operand angle and x and y are the operands stored in appropriate floating point registers.

The approximation tables for computation of the above functions are stored in Rom tables which also contains other constant

MMX (Multimedia extension):-

Intel introduced the MMX (Multimedia extension) technology at a time when there was a tremendous need to improve 2-D and 3-D imaging for multimedia applications.

Most of the algorithms in multimedia applications involve operations on several pixel(picture cell) simultaneously. Viz in a colour image , a pixel comprises 3-copmponents- red, green and blue where each component of the pixel varies from 0 to 255 in each component. In case of a black and white image, a pixel may be represented by 18-bit number.

Most of the image processing algorithms and image compressions technique required for multimedia applications involve matrix multiplication and matrix convolution type of operations and involve operations on multiple number of pixel simultaneously.

Thus most of the multimedia application require SIMD(Single Instruction stream Multiple data stream) kind of architecture. Intel provides a set of 57 MMX instructions, these instruction help the programmer to write efficient programs for image filtering, image enhancement, coding and other algorithms.

ABOUT THE AUTHORS

Dr Vilas M Ghodki is an Associate Professor in the Department of Electronics at Shiksha Mandal's Jankidevi Bajaj College of Science, Wardha, Maharashtara (India). He did his Ph. D. in subject Electronics from RTM Nagpur University, Nagpur. His area of specialization is Virtual Instrumentation and Programming. (Contact email - vilasghodki@rediffmail.com)

Dr Satish J Sharma is Head, Department of Electronics and Computer Science, Post Graduate Teaching Department, RTM Nagpur University, Nagpur. He has number of research papers published in subject of Electronics. Number of students completed Ph. D under his guidance. (Contact email – sharmasat@gmail.com)

Mrs. Trupti A Dange is an Assistant Professor in Dr Ambedkar collge, Nagpur. She is persuing her Ph D in the field of Virtual Instrumentation. She has been teaching M Sc, B Sc, B C A in subjects Elelctronics and Computer Science. (Contact email-ashishdange1973@gmail.com)

www.ingramcontent.com/pod-product-compliance
Lightning Source LLC
Chambersburg PA
CBHW051653170526
45167CB00001B/442